Field Guide to

Binoculars and Scopes

Paul R. Yoder, Jr.
Daniel Vukobratovich

TO MY OLD FREIND DAVE TWIEG

SPIE Field Guides
Volume FG19

John E. Greivenkamp, Series Editor

MAY THIS HELP YOU SEE FARTHER!

SPIE
PRESS

Bellingham, Washington USA

Daniel Vukobratovich

Library of Congress Cataloging-in-Publication Data

Yoder, Paul R.
 Field guide to binoculars and scopes / Paul R. Yoder and
Daniel Vukobratovich.
    p. cm. – (The field guide series ; FG19)
 Includes bibliographical references and index.
 ISBN 978-0-8194-8649-3
 1. Binoculars. 2. Telescopes. I. Vukobratovich, Daniel. II. Title.
 QC373.B55Y63 2011
 681'.412–dc22

                                                    2011009994

Published by

SPIE
P.O. Box 10
Bellingham, Washington 98227-0010 USA
Phone: +1.360. 676.3290
Fax: +1.360.647.1445
Email: books@spie.org
Web: http://spie.org

First printing

Printed in the United States of America.

Introduction to the Series

Welcome to the *SPIE Field Guides*—a series of publications written directly for the practicing engineer or scientist. Many textbooks and professional reference books cover optical principles and techniques in depth. The aim of the *SPIE Field Guides* is to distill this information, providing readers with a handy desk or briefcase reference that provides basic, essential information about optical principles, techniques, or phenomena, including definitions and descriptions, key equations, illustrations, application examples, design considerations, and additional resources. A significant effort will be made to provide a consistent notation and style between volumes in the series.

Each *SPIE Field Guide* addresses a major field of optical science and technology. The concept of these Field Guides is a format-intensive presentation based on figures and equations supplemented by concise explanations. In most cases, this modular approach places a single topic on a page, and provides full coverage of that topic on that page. Highlights, insights, and rules of thumb are displayed in sidebars to the main text. The appendices at the end of each Field Guide provide additional information such as related material outside the main scope of the volume, key mathematical relationships, and alternative methods. While complete in their coverage, the concise presentation may not be appropriate for those new to the field.

The *SPIE Field Guides* are intended to be living documents. The modular page-based presentation format allows them to be easily updated and expanded. We are interested in your suggestions for new Field Guide topics as well as what material should be added to an individual volume to make these Field Guides more useful to you. Please contact us at fieldguides@SPIE.org.

John E. Greivenkamp, *Series Editor*
College of Optical Sciences
The University of Arizona

The Field Guide Series

Field Guide to Geometrical Optics, John E. Greivenkamp (FG01)

Field Guide to Atmospheric Optics, Larry C. Andrews (FG02)

Field Guide to Adaptive Optics, Robert K. Tyson & Benjamin W. Frazier (FG03)

Field Guide to Visual and Ophthalmic Optics, Jim Schwiegerling (FG04)

Field Guide to Polarization, Edward Collett (FG05)

Field Guide to Optical Lithography, Chris A. Mack (FG06)

Field Guide to Optical Thin Films, Ronald R. Willey (FG07)

Field Guide to Spectroscopy, David W. Ball (FG08)

Field Guide to Infrared Systems, Arnold Daniels (FG09)

Field Guide to Interferometric Optical Testing, Eric P. Goodwin & James C. Wyant (FG10)

Field Guide to Illumination, Angelo V. Arecchi; Tahar Messadi; R. John Koshel (FG11)

Field Guide to Lasers, Rüdiger Paschotta (FG12)

Field Guide to Microscopy, Tomasz S. Tkaczyk (FG13)

Field Guide to Laser Pulse Generation, Rüdiger Paschotta (FG14)

Field Guide to Infrared Systems, Detectors, and FPAs, Second Edition, Arnold Daniels (FG15)

Field Guide to Laser Fiber Technology, Rüdiger Paschotta (FG16)

Field Guide to Wave Optics, Dan Smith (FG17)

Field Guide to Special Functions for Engineers, Larry C. Andrews (FG18)

Field Guide to Binoculars and Scopes, Paul R. Yoder, Jr. & Daniel Vukobratovich (FG19)

Field Guide to Binoculars and Scopes

The intent of this *Field Guide* is to explain the functions and configurations of various types of binoculars and scopes to the beginner as well as to the experienced user. We also attempt to show *why* a given instrument is designed the way it is.

Binoculars of various sizes—ranging from pocket size to giant models, high magnification and wide angle types, and ones used for military, law enforcement, marine and amateur astronomical applications—are considered. Scopes include small monoculars, spotting scopes, riflescopes, weapon sights, and astronomical types as large as 300 mm. Mounts for the larger instruments are also considered. Theoretical explanations of optical and mechanical systems performance are summarized.

We acknowledge with thanks Bushnell Outdoor Products, Carl Zeiss AG, Carl Zeiss Sport Optics, Leuopold & Stevens, Möller-Wedel GmbH, Questar, Schultz Loupe Direct, Steiner, Swarovski Optik KG, and the University of Arizona's College of Optical Sciences for technical information and illustrations included here.

We also thank John Greivenkamp, Wright Scidmore, and Bruce Walker for reviewing the manuscript and offering valuable suggestions for corrections and clarifications.

Any mention of specific hardware in this *Field Guide* is not meant to be an endorsement, but rather, it is intended to cite an example of a certain instrument configuration or design feature of potential interest to the reader.

The authors dedicate this *Field Guide* with love to the memory of Paul's late wife, Betty, and to Daniel's wife, Suzanne.

Paul R. Yoder, Jr.
Norwalk, Connecticut

Daniel Vukobratovich
Tucson, Arizona

Table of Contents

Table of Contents

Table of Contents

Glossary of Symbols

A	Age, distance, prism face width
A/R	Antireflection (coating)
AFOV	Apparent field of view
AIM	Aerial image modulation
AS	Aperture stop
B	Stereo baseline
BFD	Back focal distance
CCD	Charge-coupled device
cd	Candela
CF	Center focus
CED	Clear eye distance
C_n^2	Index of refraction structure
D	Diopter (unit)
D_{EP}	Diameter of entrance pupil
D_{EYE}	Diameter of eye pupil
D_{FS}	Diameter of field stop
D_{OBS}	Diameter of obscuration
D_{XP}	Diameter of exit pupil
e	Naperian logarithm base
E	Elastic modulus, efficiency
EFL	Effective focal length
EP	Entrance pupil
ER	Eye relief
f_{EP}	EFL of eyepiece
f_{OBJ}	EFL of objective
f_n	Fundamental vibrational frequency
f/number	Relative aperture
FOV	Field of view
GEM	German equatorial mount
GOTO	Go to (drive; mount)
I	Moment of inertia
I_C	Critical angle of incidence
IF	Internal focus
IP	Inverted Porro
IPD	Interpupillary distance
L	Distance, luminance level
LCD	Liquid crystal display
LED	Light-emitting diode
LOS	Line of sight
lp	Line pair

Glossary of Symbols

M	Magnification
MgF_2	Magnesium fluoride (A/R coating)
M_L	Limiting magnitude
M_V	Apparent visual magnitude
mil	US military angular unit
MLD	Multilayer dielectric (A/R coating)
MTF	Modulation transfer function
n	Refractive index
NIR	Near infrared
O	Axis offset
R_{EYE}	Resolution of eye; detection range of eye
RFOV	Real field of view
R_{OPT}	Resolution of the eye through a optical instrument, detection range of the eye through an optical instrument
RFT	Richest-field telescope
r_0	Fried parameter
R_{OPT}	Resolution with optics; detection range with optics
R_V	Visual range
S	Distance, Strehl ratio
S_{OEA}	Strehl ratio due to obscuration
t	Axial path length, time
T	Temperature, light transmission
TIR	Total internal reflection
T_s	Settling time
XP	Exit pupil
XPD	Exit pupil distance
V_{EYE}	Visual acuity of eye
V_{OPT}	Visual acuity with optics
VTR	Vapor transmission rate
W	Mass flow of water
α	1/2 real field of view in object space
β	1/2 apparent field of view in image space
Δ	Difference between parameters; eyepiece focus motion per diopter
ε	Ratio of obscuration diameter to D_{EP}
η	Damping coefficient
θ	Angle designation
λ	Wavelength
ρ	Density
τ	Absorption coefficient of glass

What Are Binoculars and Scopes?

Binoculars and scopes are **afocal** optical instruments. Their objects and images are nominally at infinity. The eyes can focus on such images.

A **binocular** is used for viewing a distant object simultaneously with both eyes. A **scope** serves the same function but uses only one eye and is also called a **monocular**. These photographs show general configurations of such instruments.

With either type, the image appears larger or **magnified**, as compared to the image seen by the unaided eye. This allows (1) finer details of the object to be resolved at a given distance or (2) given size details to be resolved at a greater distance.

A binocular has two scopes attached by a hinge or an equivalent mechanism that allows the eyepiece axes to be separated by a user-variable distance equal to the **interpupillary distance** (**IPD**) of the user's eyes. The optical axes of these scopes are nominally parallel.

With a binocular, the images are dissimilar because the object is seen from very slightly different angles. The separation between the objective axes is called the **stereo baseline**. Most users are able to fuse these dissimilar images in their brains to produce a **stereoscopic image**. Within limits, such an image allows perception of depth between objects at different distances from the observer.

How Are These Instruments Used?

Binoculars and scopes are typically used for:

- General observation of the region surrounding the user from a vantage point or while bird watching, hiking, mountain climbing, etc.

- Location, observation, identification, and tracking of animals, and/or marine life while hunting, recreational fishing, or commercial fishing.

- Observation of participants, scenery, and pageantry in theatrical presentations, concerts, parades, events, etc.

- Observing indoor and outdoor sports events.

- Accuracy-of-fire evaluation during firearm target practice.

- Measuring distances and azimuthal directions to objects in marine environments using binoculars equipped with rangefinders, integral compass displays, and/or reticles calibrated for angular measurement.

- Inspection and remote damage assessment of structures such as bridges, buildings, power lines, and roads following ice storms, floods, tornados, etc.

- Ground level, nautical, and/or aerial surveillance during search and rescue operations necessitated by avalanches, hurricanes, and other disasters.

- Detecting and observing military targets on the battlefield, as well as directing weapons fire onto hostile targets with binoculars and scopes specially designed for such activities.

- Observation of celestial objects such as the moon, nearby planets, comets, and stellar objects.

Basic Optical System Parameters

In an afocal system, **magnification** M relates the angular size of the image-space total **apparent field of view** (**AFOV**) to the object-space total **real field of view** (**RFOV**). The sketch below shows three parallel input rays coming from one off-axis point in an infinitely distant object at an angle α to the axis. The image-erecting subsystem (comprising prisms or lenses) turns these rays over in both directions. To first order, the rays exit parallel to each other and at the angle β to the axis.

The system's **aperture stop** (**AS**) is the physical hole that limits the size of the beam from an axial point object. Usually, it is the inside diameter of a mechanical part, such as the cell holding the objective lens. From the front, that hole (or its image if the AS is internal) can be seen directly. When seen from object space, this is called the **entrance pupil** (**EP**). From image space, it is called the **exit pupil** (**XP**). The **principal**, or **chief**, ray enters at angle α aimed toward the center of the EP and exits at the angle β, nominally passing through the center of the XP.

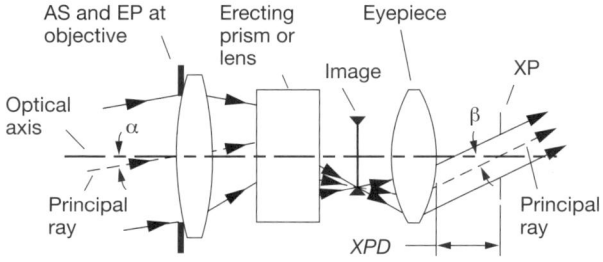

The system magnification may be expressed as $M = \beta/\alpha$ or as $M = \tan\beta/\tan\alpha$. The angles are in degrees. The latter equation is used in this book for terrestrial instrument applications.

If the image erecting means does not introduce magnification, M also can be expressed as $M = f_{OBJ}/f_{EP}$ in terms of the lens **effective focal lengths** (**EFLs**).

Basic Optical System Parameters (cont.)

M is abbreviated as a number followed by "×." A magnification of 7 times would then be written 7×.

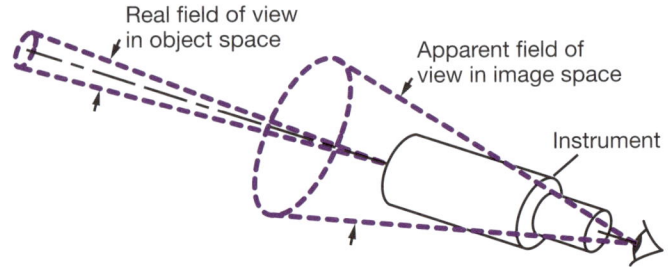

The RFOV is frequently specified as its width in meters measured at 1000 m. Hence, a total angular field of $2\alpha = 7.26°$ equals $(2)(\tan\alpha)(1000) = 127$ m at 1 km.

Looking at the eyepiece of the instrument when held ~30 cm in front of the eye and pointed toward a bright scene, such as the daytime sky (not the sun), a circular concentration of light called the exit pupil (XP) can be seen. Its diameter is $D_{XP} = D_{EP}/M$. The XP is at an **exit pupil distance** (**XPD**) beyond the last lens surface of the eyepiece. This distance is also frequently called the **eye relief** (**ER**). The principal (or chief) ray of the transmitted beam crosses the axis at the AS, EP, and XP.

Military binoculars and scopes are often designed to locate the XP midway between the eye's pupil and its center of rotation, or typically ~6.3 mm beyond the cornea. This helps maximize access to the AFOV as the eye pupil moves off axis while the eye scans the field. Aberration of the XP location is usually neglected for simplicity.

> Caution is suggested in using supplier data for AFOV, because the way it relates to RFOV through M is usually not specified.

Instrument Size and Weight

Design of a scope or a binocular is a compromise between:

(1) a desire for large optical apertures to capture a maximum amount of light from the object and funnel it into the user's eye(s), and

(2) a desire for compact packaging and minimum weight.

Weight depends on the total volumes of the materials used and their densities. The materials for lenses and prisms are optical glasses selected for optical performance reasons rather than low density. Metals, plastics, or fiber-reinforced plastics are used for housings, cells, etc. Plastics weigh less than metals, but they are less durable. Many instruments have resilient protective coverings, such as rubber, that are useful, but they add weight and bulk. Binocular weight is discussed on pages 90 and 91.

Weight should be minimized because the instruments are often manually transported, and many are held by hand during use. While weight contributes inertial resistance to applied forces and tends to steady the line of sight, it also causes muscular fatigue that progressively reduces steadiness and limits duration of use.

> As a rule of thumb, the practical upper limit for weight of a handheld scope or binocular is about 2 kg (4.4 lb).

The shapes of handheld instruments should be **ergonomic** for easy handling and have weight distribution such that support is provided at or near the assembly's **center of gravity**. This minimizes angular moments that must be counteracted by muscular forces that tire the user.

These discussions are limited to scopes with apertures <300 mm. Large scopes and binoculars weigh many kilograms and must be supported on a stable structure or mount (see pages 40 to 47).

Structure of the Eye

The human eye is biologically complex but simple as an optical system. As shown in the top-sectional view of a right eye below, it has two image-forming elements: the **cornea** and the **crystalline lens**. Images are formed on the concave **retina**, which has an array of photoreceptors (cones and rods) covering its surface. These are connected electrochemically to the brain through optic nerve fibers.

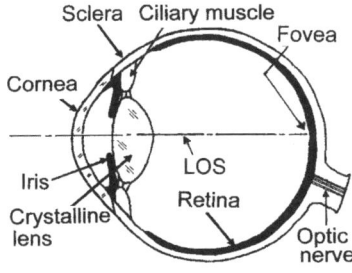

The **iris** changes aperture size as a function of the scene luminance while the lens changes its shape by action of the ciliary muscle attached to its rim. This change in shape changes the focal length of the cornea/lens combination and focuses the image of interest on the retina.

When the brain selects a point of interest on the observed object, the eyeball rotates in its socket so that its **line of sight** (**LOS**) connects the object point to the center of the entrance pupil (image of the iris as seen through the cornea). This LOS continues to the center of the **fovea**, which is a small retinal area where the photoreceptors (cones) are the smallest (~2.5 μm) and most closely packed, hence providing the sharpest vision. Cones provide **photopic** vision at higher scene luminances and in color, while rods provide somewhat monochromatic **scotopic** vision at lower scene luminances. See page 14 for more details.

Outside the fovea, the cones become progressively sparser and larger (to ~10 μm). With distance from the fovea, rod concentration increases, and the rods grow from ~3.0 μm to ~5.5 μm. At the rim of the retina, only rods are present, so **peripheral vision** is degraded but still adequate to provide cues to the brain regarding motions of objects or light intensity changes that might represent objects of interest for more detailed examination.

Pupil Size

The diameter D_{EYE} of the entrance pupil of the human eye depends largely on the **luminance level** L of the observed scene. For an average young adult eye, after a suitable time for adaptation following a change in luminance, the approximate relationship is:

$$\log D_{EYE} = 0.8558 - 0.0004[\log(0.3142L) + 8.1]^3$$

The graph below shows the variation of D_{EYE} as L changes from daylight through maritime twilight to night. The widely accepted values for average pupil size are ~2 mm in daylight and ~7 mm when fully dark adapted.

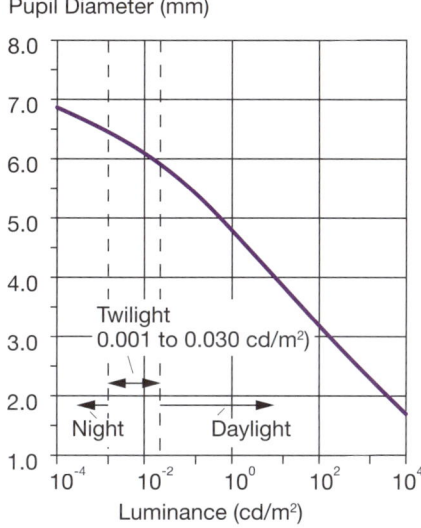

The dark-adapted pupil diameter varies significantly with age A of the individual per this empirical relationship:

$$D_{EYE} = 9.08 - 0.082A + 0.00037A^2$$

Pupil Size (cont.)

The relationship between dark-adapted pupil diameter and age is shown below. A significant variation in D_{EYE} exists among individuals of the same age. This is not reflected in this graph.

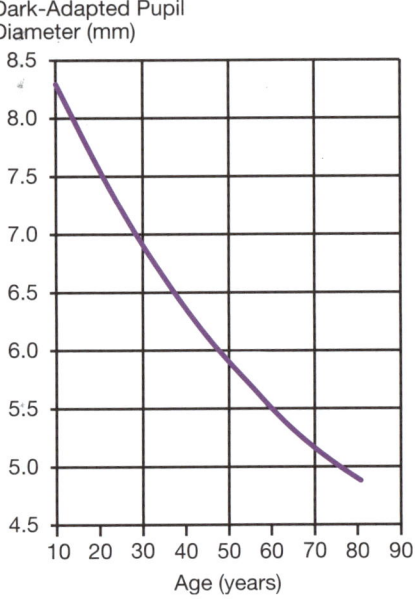

If D_{EYE} is smaller than D_{XP}, the eye becomes the system XP. For example, if D_{EYE} is 5 mm, a system with $D_{EP} = 50$ mm and $D_{XP} = 7.1$ mm is stopped down by a factor of 5/7.1 to a 35.2-mm aperture. A large portion of the system EP's aperture is then unused. Some aberrations may then be reduced so that optical performance may improve slightly.

If, at low light level, D_{EYE} opens to the system's D_{XP}, the full EP aperture is used and aberrations are not reduced. The eye's resolution capability is reduced because its pupil is larger so that the performance of the optics/eye system may be acceptable. This is an important design principle for large-aperture astronomical binoculars

Interpupillary Distance

Interpupillary distances (IPDs) of adults typically range from ~52 mm to ~75 mm. It is important for the corresponding separation of the exit pupils of binoculars to be adjustable over at least that same range. Most binoculars intended for use in the applications discussed here are hinged to satisfy this requirement.

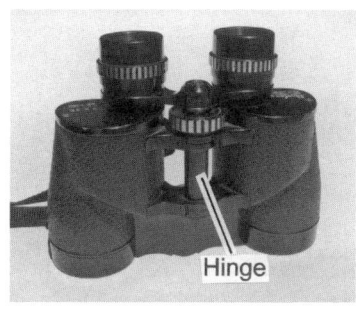

Children who are learning to use binoculars may have IPDs as small as perhaps 46 mm. Some binoculars with double hinges provide IPDs smaller than this and would be suitable for use by children.

In motion pictures or on television, the view through a binocular may be shown as two partially overlapping circles. This is not the proper condition for aligning a binocular to the eyes because the IPD setting of the instrument is incorrect. When proper alignment is achieved, the left and right images combine to provide comfortable binocular vision.

Most binoculars have IPD scales that allow a predetermined setting of eyepiece separation to be chosen by each user. The **scale** is attached to one side of the binocular, while the **index** is provided on the other side of the instrument.

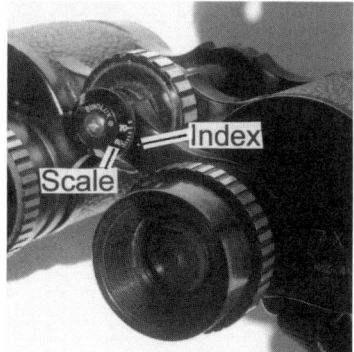

Resolving Power

The ability of the eye to **resolve** details in an object viewed unaided or through a scope or binocular varies significantly with **luminance** L of the scene observed. Typical values for the unaided eye are listed here.

Condition	Luminance (cd/m^2)	Unaided-Eye Resolution
Daylight	>0.03	~4 arcmin to ~40 arcsec
Twilight	0.03 to 0.001	~4 to ~13 arcmin
Night	<0.001	~13 to ~30 arcmin

This variation is plotted below. Values are averages based on testing. Variations from one individual to another are ~20%.

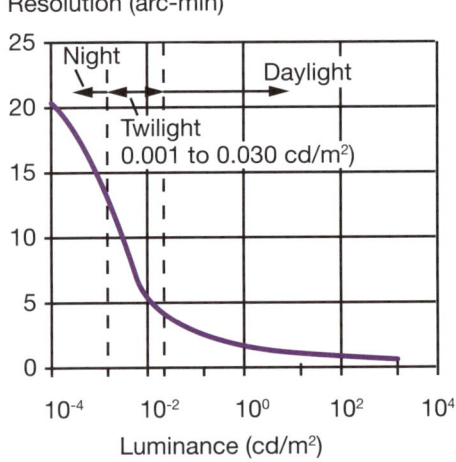

The resolution capability of the unaided eye at the center of its field of view in bright daylight is normally assumed to average at ~1 arcmin. This means that it can just resolve details in a high-contrast, black-line/white-space object subtending ~2 arcmin.

Resolving Power (cont.)

Such an object, usually referred to as a "**line pair**" (lp), is widely used as one element of a **resolution chart** for testing the eye unaided or with magnification provided by an optical instrument.

The unaided eye's resolution of 2 arcmin/lp is equivalent to ~58 mm/lp at 100-m distance. With a perfect 10× binocular or scope, details measuring ~5.8 mm/lp should then be resolved. That resolution may be adequate to identify a bird at that distance by its shape, but it may not reveal intricate feather shapes, especially if the image quality is not perfect or the atmosphere is not clear and still.

When trying to read numbers or letters, it is generally necessary for each character to subtend at least five elements or 2.5 lp. The Snellen "**E**" on an optometrist's eye chart is designed on this basis. One horizontal line and one horizontal space constitute an element. If the character subtends more elements, the recognition process is easier.

Two typical related tasks in military applications that involve resolving power of the eye are (1) to recognize a distant object seen on the battlefield as an armored vehicle (tank) rather than a truck and (2) to identify it as a particular model of tank to determine if it poses a potential threat. The angular subtense at the eye of the object's image should comprise at least four elements for task (1) and at least seven elements for task (2).

The resolving power of the eye is reduced if the observed object has inherently low **contrast**, as is the case for most natural objects. Reductions in contrast also occur because of lighting irregularities, atmospheric effects, diffraction, aberrations in the eye, focus errors, and effects of misalignments and aberrations in any optical system employed.

Accommodation

The normal eye can focus on objects located at various distances. This change results as the ciliary muscles (which are relaxed when viewing an infinitely distant object) contract and steepen the surface radii of the crystalline lens, shortening the lens focal length so that nearer objects are focused on the retina. Objects at only one distance are in sharp focus at any one time. Shifting focus is involuntary and results when the brain shifts attention from an object at one distance to an object at some other distance. This process is called **accommodation**.

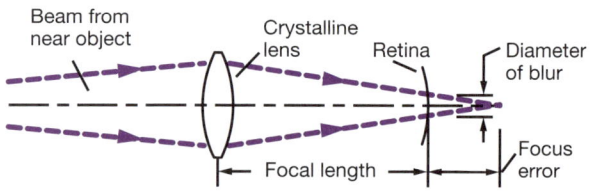

It is customary to express focal length, focus error, and accommodation in diopters, where one diopter corresponds to the reciprocal of the focal length of a lens with a focal length of 1 m. The focal length of the average normal eye is ~17.1 mm or ~58.5 diopters.

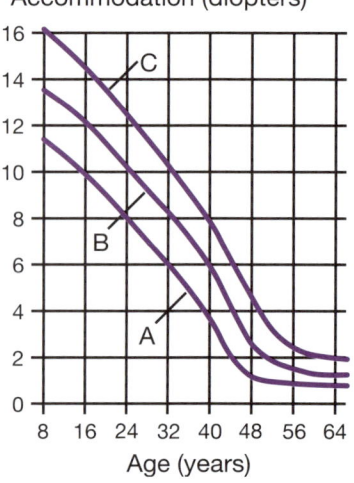

Accommodation varies with age as shown here. Curves A and C are extreme measured values for a selected population. Curve B shows the average of A and C.

Focus errors as small as ~0.2 to ~0.5 diopters can be detected by the brain and lead the eye to accommodate within a few seconds.

Stereoscopic Capability

When two objects located at different distances from the observer are seen with unaided eyes, disparities exist between the retinal images. In the top view below, the angles θ_L and θ_R to objects O_1 and O_2 differ slightly because of the distance $B = IPD$ separating the eyes laterally, where B is the stereo baseline.

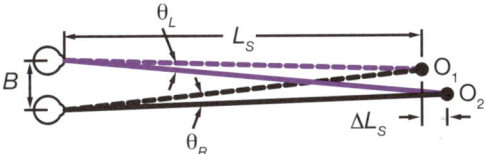

The ability of the eye to distinguish small differences $\Delta\theta$ between θ_L and θ_R is called **stereo acuity** and typically ranges from ~10 to ~30 arcsec. About 10% of the adult human population lacks this capability. The minimum detectable axial separation of distant objects and the maximum distance for **stereo vision** are approximated as

$$\Delta L = 1000 L_S^2 \Delta\theta/B \ \text{ and } \ L_{MAX} = B/(1000\Delta\theta).$$

Here, L_s and ΔL_s are distance in m, B is in mm, and $\Delta\theta$ is in radians.

> For example, let $B = 65$ mm, $\Delta\theta = 30$ arcsec (1.45×10^{-4} rad), and $L_s = 40, 80, 160,$ and 320 m. Then, $L_{MAX} = 448$ m and $\Delta L_s = 4, 14, 57,$ and 228 m, respectively. The ability to discern the separation between objects decreases with L_s and breaks down when ΔL_s is a large fraction of L_s.

At long distances, **parallax** due to object lateral motion relative to its surroundings, **atmospheric effects**, **perspective**, and relative size cues become more effective in judging relative distance than does the stereoscopic effect.

A binocular extends the range of stereo vision by virtue of its magnification and, if it has prisms that offset the objective/eyepiece axes, by increasing the stereo baseline. This topic is discussed on page 48.

Luminosity and Chromatic Sensitivities

The eye cannot accurately judge the absolute **luminosity** of an object or scene, but it can detect small differences in luminosity between two adjacent areas in the observed scene. This helps the observer discern details in complex scenes such as those that occur in bird watching, hunting, or camouflage penetration.

The eye is also unable to accurately judge absolute colors of objects within a scene, but it is very good at matching colors of nearly the same wavelength if the areas bearing those colors are adjacent.

The **relative sensitivity** of the eye varies with the color, i.e., wavelength, of the scene. The following graphs depict that variation over the spectral range characteristic of the visible spectrum (left) and over the near-infrared (NIR) spectrum (right). The peak sensitivity of the left curve (called the **photopic response**) in daylight is at ~0.55 μm. When the eye is dark adapted, the shape of the left curve (then called the **scotopic response**) stays about the same, but its peak shifts left to ~0.51 μm.

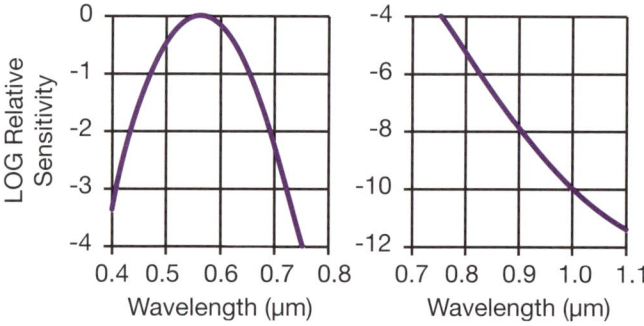

The NIR graph is useful for estimating the visibility of scenes illuminated by near-infrared searchlights.

Galilean Systems

The oldest and simplest type of scope was described by **Lippershey** in a 1608 patent application. **Galileo** used this type of scope in his early astronomical studies, and it has since borne his name.

The system has only two lenses: an objective lens (with positive EFL) and an eyepiece (with negative EFL). In the schematic below the **marginal ray** (ray 1) enters and exits parallel to the axis, indicating this to be an afocal telescope. An erect image is formed without needing prisms. This fact is indicated by the principal or chief ray (ray 2) that enters traveling upward and continues traveling upward as it enters the eye.

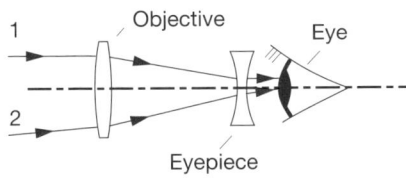

Prior to the invention of prismatic binoculars in 1893, binoculars consisted of two Galilean scopes attached together. They were generically named **field glasses**; a small field glass is shown here. It is called an **"opera glass"** because it is useful in the opera house or theatre; it is compact and lightweight enough to be easily carried in a pocket or purse. Magnification is low (~2× to ~3×) and the real field of view is ~5° to ~3°. The field is limited by the diameter of the objective (see ray 2 in the optical schematic above). Simplicity limits the image quality of these instruments. The eye pupil is the AS of the system. In order to see the apparent field, the eye must be very close to the eyepiece. This makes viewing difficult for eyeglass wearers.

Galilean Systems (cont.)

Field glasses with magnifications of 5× or 6× have long been used in hunting and for viewing sports events. They were used during both World Wars. With only two optical components

and four glass-to-air interfaces, light transmissions were higher than in designs using prisms. This advantage continued until **antireflection (A/R)** surface coatings were invented at the end of WWII to improve transmissions of more complex designs.

> With A/R coatings, field glasses are effective at low scene luminances because the D_{XP} always matches the D_{EYE}.

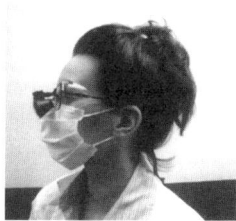

Galilean optics are also used as head-mounted, long-working-distance magnifiers (**loupes**) by surgeons and dentists. Several types are available.

Listed here are some representative WWII-era field glasses and a typical surgical/dental loupe.

Instrument	M	RFOV	Weight (g)
US Army 08	3×	4.5°	630
German Fernglas 08	6×	4.4°	480
UK Binocular V.F. 2506	2.5×	10°	800
Japanese Military	4×	6°	420
Surgical/Dental Loupe	3×	10 cm at 10 cm	71

Keplerian Systems

Scopes and binoculars based on the **Keplerian** system design have replaced the Galilean system for all but the least-demanding applications. Long used in refracting telescopes for astronomical applications, the Keplerian scope has an **aerial image** located between the objective and the eyepiece. It is located at the focal plane of the objective and serves as the object for the eyepiece. The XP here is a **real image** of the AS, whereas that of the Galilean system is actually the pupil of the eye.

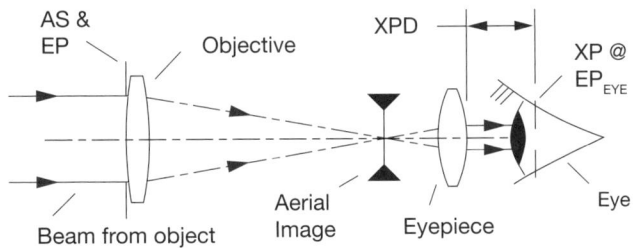

The image produced here is inverted both vertically and horizontally. For terrestrial applications, an erect image is needed. Originally, lenses were used for this purpose, as shown schematically below. Although this made the instruments very long and subject to misalignment, they were used aboard ships, on the battlefield, and by birdwatchers until prismatic erecting systems came into use. A classic example of a lens erecting scope is shown on the next page. It has three brass tubes that slide into each other and into the main housing for storage. Early binoculars were made from two of these scopes. They, too, were very long and quite difficult to hold during use. Maintaining parallelism of the lines of sight also was very difficult.

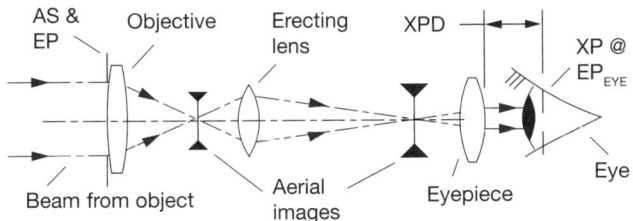

Keplerian Systems (cont.)

Much shorter scopes for use on rifles and other handheld weapons are usually provided with lens-type erectors because the optics can then have a favorable cylindrical configuration that is easily attached to a gun barrel.

Zeiss patented the use of prisms in Keplerian binoculars in 1894 and sold the first units that same year. That improvement shortened the scopes, erected the images, and improved durability. Two air-spaced Porro prisms were used in each half of such a binocular, as shown below. The "arrow crossed with a drumstick" symbols indicate how the image orientation changes. Later designs have the prism's surfaces nearer each other or cemented.

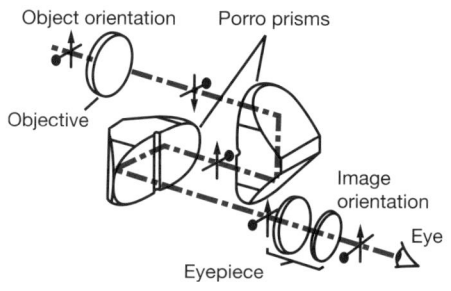

Several other types of prisms have been used to erect the images in binoculars and scopes. The most common ones use Type 1 Porro prisms (shown here) and Type 2 Porro prisms (shown on page 65). Both provide lateral offsets of the objective and eyepiece axes to increase the stereo baseline. Other designs use roof prisms to provide an in-line optical path, i.e., one with no axis offset. See page 66.

Binocular Types—General Considerations

Binoculars are made in a variety of types, each with a particular set of capabilities. General ways to distinguish one binocular from another are as follows:

- By magnification and objective lens aperture. For example, a "6 × 30" instrument has a magnification of 6 and 30-mm apertures. Such an instrument is referred to as a "6 by 30."

- By overall size. Sizes range from compact types that fit into a jacket pocket, to mid-size types for handheld use, to full-size and giant types, and to astronomical types.

- By magnification.

- By real field of view.

- By erecting prism configuration.

- By method for focusing on the target.

- By application. Grouping binoculars by application classifies them for nature study and bird watching, sports observation, scoring target shooting with small arms, hunting, marine, military, law enforcement, and astronomical uses.

Some binoculars feature special integral capabilities including variable magnification, image stabilization, and distance measurement to a target (range finding). Other models have compasses for showing the LOS direction to a target of interest relative to magnetic north. Most binoculars are compatible with a still or video camera attachment to record the observed scene. A few have integral digital cameras.

Instruments falling into the various categories listed above are described in more detail on the following pages.

Compact Binoculars

Compact binoculars have objective sizes of 15 mm to 25 mm and magnifications of 6× to 12×. Most weigh <300 g. Their XPs approximate the 2-mm to 2.5-mm daytime pupil diameter of the adult eye. The daylight efficiency of a compact binocular may then be equivalent to those of full-size units in which the pupil sizes do not match. This size of binocular has poor efficiency in twilight conditions and is unsuitable for night use. Their AFOVs typically are < 50°, and XPDs range from ~9 mm to ~16 mm. This table lists examples.

Company and model	M	Aper. (mm)	Prism type	RFOV (deg)	Weight (g)
Nikon HC	6×	15	Porro	8.0	130
Zeiss Victory	8×	20	Roof	6.8	225
Pentax Papilio	8.5×	21	Inverted Porro	7.5	284
Nikon DCF	8×	25	Inverted Porro	6.8	280
Swarovski Pocket	10×	25	Roof	5.4	230

Some compact binoculars have a short focus distance, which is useful for viewing museum exhibits and small objects. The Pentax Papilio, shown here, can focus to ~45 cm. It is a magnifier with a long working distance  for study of objects such as butterflies (hence the name from the Latin).

Mid-Size Binoculars

For daylight through the start of twilight and a binocular with typical light transmission, the adult eye pupil diameter does not exceed ~5.0 mm. The maximum useful magnification for a handheld binocular is considered to lie between ~6× and ~8×. Based on these two criteria, a **mid-size binocular** is defined as 6 × 30 to 8 × 40. The AFOVs of such binoculars typically range from ~46° to ~60°. Either Porro or roof types of erecting prisms are used.

Mid-size binoculars are popular when size and weight are more important than efficiency in low light. During daytime use, they can be identical in efficiency to a full-size binocular. During extended use, a mid-size unit is less fatiguing than a larger model.

This photo shows a mid-size Leupold Golden Ring binocular with switchable magnification of 7× and 12×.

This table lists a few examples of mid-size binoculars:

Company and model	*M*	Aper. (mm)	Prism type	RFOV	Weight (g)
Nikon Nature	8×	32	Porro	7.5°	575
Zeiss Conquest	8×	30	Roof	6.9°	495
Steiner Merlin	8×	32	Roof	6.4°	623
Leupold Golden Ring Switch	7/12×	32	Roof	7.1/4.1°	608
Bushnell Spectator	8×	40	Porro	8.1°	765

Full-Size Binoculars

Binoculars with objectives of 40-mm to 56-mm aperture are designated here as **full-size** instruments. They may be preferred over mid-size binoculars when better performance is desired under twilight conditions. Using methods shown on pages 51 and 52, the twilight efficiency of a 7×50 binocular is ~1.05 times better than that of an 8×40 instrument, while that of a 15×60 binocular is about 1.6 times better than that of the 7×50 version. Hunters and bird watchers often use the 15×60 for this reason. Tripod mounting is helpful for extended use.

Here are examples of typical full-size binoculars:

Company and model	*M*	Aper. (mm)	Prism type	RFOV	Weight (g)
Steiner Predator	12×	40	Porro	5.0°	690
Nikon Marine	7×	50	Porro	7.5°	1170
Garret Classic	7×	50	Porro	7.5°	978
Leupold Olympic	12×	50	Roof	4.8°	728
Leica Duovid	10/15×	50	Roof	5.3/3.5°	1250
Docter Nobilum	10×	50	Porro	6.8°	1400
Nikon DCF	12×	56	Roof	5.5°	1180

This photo shows the classic Docter 10×50 Nobilum, which has an RFOV of 6.8°. It is rugged, and has been found to perform well for long-distance nature study as well as for amateur astronomical viewing.

Giant Mounted Binoculars

Impressive performance can be achieved with **giant binoculars** if the instrument is mounted securely. Objective apertures here are 70 mm to 150 mm and magnifications are 10× to 40×. Some have interchangeable eyepieces, while a few have eyepiece turrets for switching from low to high magnification.

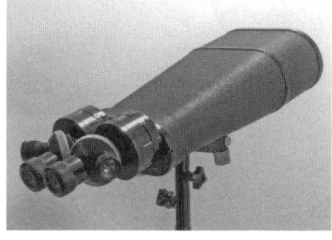

This switchable-eyepiece example is a "Border Hawk" from The People's Republic of China. It is a 25/40×100 model with a nominally horizontal LOS and limited elevation motion.

Other giant binoculars are used for such applications as amateur astronomy (including comet hunting), commercial fishing, and military observation.

The heavier models in this class must be mounted. The simplest mount is a camera tripod with pan-tilt head. This is inconvenient if the system is to be used to access elevated objects. Some instruments have inclined eyepieces to solve this problem. See pages 40 through 47 for further information about mounts. Typical giant binoculars are listed here:

Company and model	*M*	Aper. (mm)	Prism type	RFOV	Weight (g)
Garrett Gemini	15×	70	Porro	4.3°	1885
Steiner Military	15×	80	Porro	3.7°	1588
Fujinon MT	15×	80	Porro	4.0°	7060*
Fujinon MTM	25×	150	Porro	2.7°	40,000*
Chinese Border Hawk	25/40×	100	Porro	2.5°/1.5°	11,600

* mounted

High-Magnification and Wide-Angle Binoculars

High-magnification binoculars are commonly defined as those with $M \geq 10\times$. They are most frequently used for big game hunting, nature study, astronomy, and special military purposes. As a class, the twilight efficiencies of these instruments are generally higher than that of a 7×50 night glass. They perform best when mounted. Typical models are as follows:

Company and model	M	Aper. (mm)	Prism type	AFOV	RFOV
Leica Ultravid	12×	50	Roof	62°	5.7°
Swarovski SLC	15×	56	Roof	60°	4.4°
Docter	18×	70	Porro	61°	3.7°
Nikon Nature	15×	60	Porro	64°	4.8°
Steiner Commander	15×	80	Porro	52°	3.7°
Garrett Gemini	30×	100	Porro	58°	2.1°

Binoculars with AFOVs > 50° are frequently called **wide angle**. ISO standard 14132–1:2002 defines them as AFOV > 60°. Larger fields produce a feeling of immersion in the target space. Some wide-angle binoculars have excessive distortion and short XPDs. Typical models are as follows:

Company and model	M	Aper. (mm)	Prism type	AFOV	RFOV
Zeiss Deltarem	8×	40	Porro	90°	14.2°
Sard Mk 43	6×	42	Porro	72°	13.8°
B&L Mk 41	6×	50	Porro	70°	13.3°
Leitz Amplivid	6×	24	Roof	72°	13.8°
Kronos BPWC2	7×	30	Porro	75°	12.5°
Miyauchi Binon	7×	50	Porro 2	68°	11.0°

Military and Law Enforcement Binoculars

The general configurations and functions of **military** and **law enforcement binoculars** are nearly the same. Differences between military and nonmilitary units are summarized here:

- Nonmilitary optics are focusable on targets at any distance. Military optics are usually focused at infinity.

- **Reticle** patterns are needed in most military instruments to measure angles for weapon aiming. These are located on glass plates in the eyepiece object plane. Few nonmilitary binoculars have reticles.

- Military instruments usually have individual focus eyepieces; nonmilitary ones often have center focusing.

- Most military binoculars use Porro prism erectors for enhanced stereoscopic vision. Nonmilitary ones use a variety of prism types.

- Military binoculars may have optical filters to protect the user's eyes from injury due to laser rangefinders and target designators.

- Glints from the objective can disclose user position. A special "**Killflash**" filter is used in military binoculars to reduce this danger. This device, comprising crossed honeycomb structures with thin but deep blades, is set in front of each objective to reduce off-axis reflections.

- **Durability** and **sealing** of lower-cost nonmilitary instruments are generally poor. Higher-cost nonmilitary and military versions must pass strict durability (shock and vibration) and immersion tests.

- Higher-cost nonmilitary and military binoculars are flushed with clean, dry gas (typically **nitrogen**) to remove internal moisture and are sealed after assembly is complete.

Military and Law Enforcement Binoculars (cont.)

Two primary types of handheld military binoculars are those intended for infantry use and the 7 × 50 night glass. Attributes of the latter are discussed on page 52.

Shown here is a cut-away view of a military binocular.

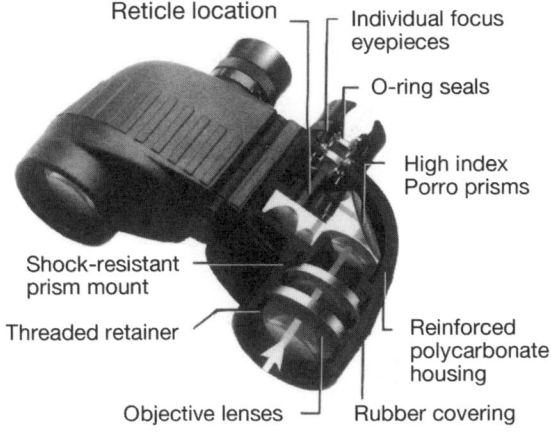

This list identifies representative military binoculars:

Origin, model, and date	*M*	Aper. (mm)	Prism type	RFOV	Weight (g)
US, M13, 1940	6×	30	Porro	8.5°	660
US, M17, 1943	7×	50	Porro	7.3°	1502
US, M19, 1966	7×	50	Porro	7.3°	970
German, EDF, 1966	7×	40	Roof	7.5°	950
UK, L12A1, 1966	7×	42	Porro	7.0°	640
German, D16, Current	8×	30	Porro	7.0°	650
US, M24, Current	7×	28	Roof	7.0°	570
US, M22, Current	7×	50	Porro	7.0°	1600

Astronomical Binoculars

Binoculars offer a low-cost, portable means for amateur astronomers to view celestial objects. They are superior to conventional astronomical scopes for visual examination of large, low-surface-brightness objects such as nebulae. They also tend to increase the visibility of faint stellar objects compared to a monocular scope of equal magnification and aperture.

Binocular efficiency when observing astronomical objects is determined first by objective size and second by magnification. For the purposes of this *Field Guide*, the ranges of aperture size and magnifications of astronomical instruments under consideration are limited to 50 mm to 150 mm and 7× to 25×. Additional concerns are eye relief, field of view, scattered light suppression, and optical correction.

Three types of binoculars are used for astronomy:

- Night glasses with a 7-mm XP. These are used to observe extended low-surface-brightness objects such as the North American Nebula (NGC 7000); effective use requires dark sky conditions and a fully dark-adapted eye. Examples include the 7×50 and 10×70.

- General-purpose astronomical binoculars with a magnification between 7× and 10×, and a pupil between 4 mm and 5 mm. These are most effective under average sky conditions and for observers without complete dark adaptation. Hundreds of deep-sky objects are visible with a 50-mm aperture binocular. A good candidate is a wide-angle 10×50 Porro prism binocular (see page 22), which is near the limit of efficiency for astronomy when handheld.

- Mounted binoculars with a magnification >15 and an exit pupil between 4 mm and 5 mm. All of the objects in the Messier catalog are visible with an 80-mm binocular. Examples include the $14 \times 70, 15 \times 80$, and 25×100.

French astronomer Charles Messier (1730–1817) cataloged 103 star clusters and nebulae.

Astronomical Binoculars (cont.)

Individual-focus eyepieces are acceptable for astronomical binoculars, since the focus is normally set at infinity. A long eye relief of at least 15 mm and ideally of 20 mm is desirable to allow the user to wear eyeglasses. An AFOV of at least 60° gives a sense of immersion into the observed scene.

Good stray light control is important in the instrument because there is often a large brightness ratio in celestial objects. Astronomical binoculars should be waterproof to minimize intrusion of moisture during dewing conditions. A tripod is used to minimize image motion and fatigue in long observing sessions.

This photo illustrates a 25×100 astronomical binocular.

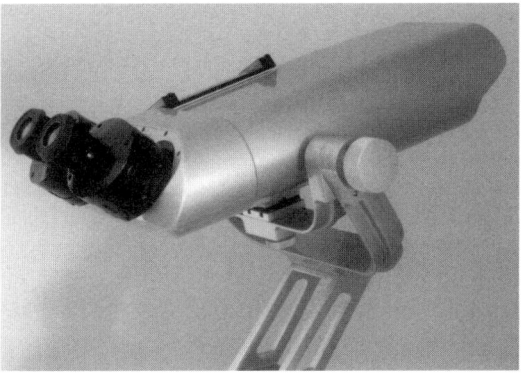

The following table lists typical astronomical binoculars:

Company and model	*M*	*D* (mm)	RFOV	Wgt. (kg)
Fujinon Polaris FMTR-SX	10×	50	6.5°	1.5
Docter Nobelim BGA	8×	56	6.3°	1.4
Nikon HP WP	10×	70	5.1°	2.0
Steiner Military	15×	80	3.7°	1.6
Oberwerk BT-100-45 (45°)	25×	100	2.5°	11.4
Fujinon MT-SX	25×	150	2.8°	18.5

Monoculars

Instrumentation collectively called **monoculars** includes any scope with only one complete optical system, as opposed to a binocular with two such systems. Some monoculars are pocket size, while others have magnifications and apertures equivalent to low-end mid-size binoculars. Early ones were Galilean, whereas modern ones have erecting prisms. Two typical pocket-size monoculars are shown here. The one on the left unfolds flat for storage. It is a recent model based on the Zeiss Jena "Turmon" design of 1921.

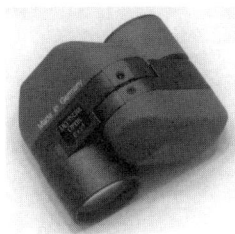

Although monoculars lack stereoscopic capability and have been found by some users to be less comfortable to use over extended periods than a binocular, their advantages in overall size, weight, and cost make them popular alternatives for many applications.

The borderline between larger monoculars and spotting scopes is indistinct. Most monoculars are constructed for low weight rather than durability. Sealing provisions are minimal. The following table lists a few small models.

Company and model	*M*	Aper. (mm)	RFOV	Wgt. (g)
Zeiss Conquest	4×	12	10.3°	45
Minox Minoscope	8×	25	6.5°	153
Brunton Eterna	6×	30	6.5°	326
Deutsche Optik Turmon	8×	21	6.3°	80
Celestron	8×	25	8.7°	99

Spotting Scopes

Spotting scopes are larger and perform better than small monoculars. They generally have many of the features of corresponding-size binoculars and cost nearly as much. Magnifications and apertures typically range from about $10\times$ to $60\times$ and 30 mm to 100 mm, respectively.

Spotting scope applications include nature study; scouting game while hunting; observing the accuracy of fire on a target range; general observation in military and law enforcement applications; serving as a small astronomical scope; or functioning as an auxiliary pointing aid for a large astronomical scope.

The spotting scope performs best if mounted on a tripod or some other rigid support with a **pan/tilt head** to allow the LOS to be aimed in a chosen direction. Some have collinear optics while others have inclined eyepieces. Some models have interchangeable eyepieces or multiple eyepieces in a turret to vary the magnification in fixed steps. Others employ zoom objectives or eyepieces for a continuous variation of magnification.

Two models are shown here. One is a 65-mm aperture refracting Swarovski scope that uses interchangeable $20\times < M < 60\times$ or $25\times < M < 50\times$ zoom eyepieces. The other is a Zeiss 60-mm aperture model with a Schmidt–Gregorian objective and fixed $30\times$ eyepiece. It is shown mounted on a heavy tripod with a pan/tilt head.

Spotting Scopes (cont.)

This figure shows the interior construction of a Swarovski spotting scope, which has interchangeable eyepieces and a sealing window to maintain the sealing of the main housing when the eyepiece is removed. Another useful feature is the rotatable tripod attachment ring. It allows the scope to be rotated about its axis to bring the eyepiece into a convenient location on a tripod.

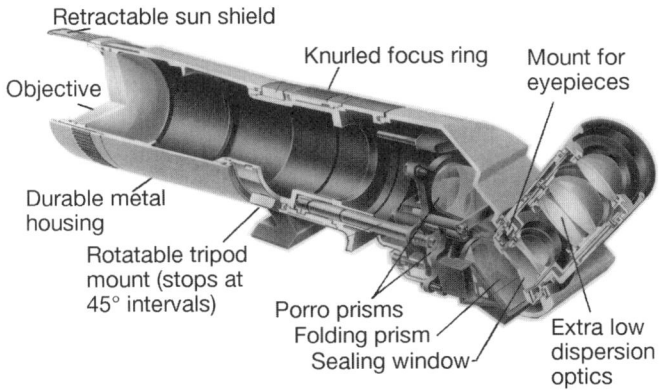

Retractable sun shield

Knurled focus ring

Mount for eyepieces

Objective

Durable metal housing

Rotatable tripod mount (stops at 45° intervals)

Porro prisms
Folding prism
Sealing window

Extra low dispersion optics

The following is a list of typical spotting scopes:

Company and model	*M*	Aper. (mm)	RFOV	Weight (g)
Leupold Compact	10×–20×	40	3.8–2.6°	448
Brunton Eterna	20×–45×	62	1.8–1.1°	190
Nikon Prostaff	16×–48×	65	1.9–0.8°	920
Kowa TSN-883	20×–60×	88	–	1520
Swarovski STM-65	20×–60×	65	2.1°	1020 (body only)

Riflescopes

Small arms, such as rifles, shotguns, and handguns, traditionally have **iron sights** as aiming references. These devices, as simple as a blade at the front and a notch at the rear of the gun barrel's top surface, provide remarkably good accuracy (better than ~0.6 m at 1000 m in the hands of an experienced shooter during rifle matches). Iron sights have long been used for hunting with rifles and in military/law-enforcement applications.

Adding a scope to the weapon provides these advantages:

- The time to fire is reduced because the aiming point (**crosshair** or reticle) and the target image are in, or can be adjusted to, the same plane of focus. Eye accommodation change is not required to bring these into alignment, and parallax is minimized.

- The target can be acquired and identified at lower scene luminances than with iron sights.

- Magnification allows more accurate hit point location. Some riflescopes have zoom adjustment so that magnification can be optimized for the conditions at hand.

The inevitable increases in cost and weight when a **riflescope** is added to the weapon are offset by improvements in performance.

A typical riflescope of contemporary design is shown here mounted to the top of a hunting rifle.

Riflescopes (cont.)

Riflescopes may have fixed or variable (zoom) magnification. The latter typically range from 1.5–5 × 20 to 8.5–25 × 56. Length and weight of a typical midsize riflescope are about 300 mm and 330 g. These parameters vary significantly with objective size and magnification.

Built-in precision azimuth and elevation mechanisms allow adjustment ("**zeroing**") of the LOS of the optics relative to the rifle bore to correct for ballistic drop and windage effects on the bullet at the target range.

Nearly all riflescopes are of Keplerian form with lens erecting systems. This form is shown on page 17. The reticle is located at one of the internal images. It then appears to the eye to be superimposed on the image of the target. The internal configuration of a typical riflescope is shown below. The central cylindrical housing is usually 25.4 mm or 30 mm in diameter and fits into rings securely attached to the gun barrel.

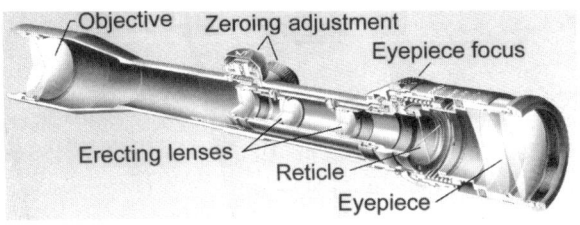

The user's eye should be located so that its entrance pupil (i.e., iris opening) coincides with the scope's XP. The latter pupil is usually located ~75 mm to ~100 mm from the eyepiece to prevent injury to the shooter's eye or face due to recoil.

If the eye is not properly aligned to the exit pupil, a portion of the magnified field of view will not be seen. Exit pupil diameter depends on magnification and objective aperture. It ranges from ~2 mm to >13 mm in common riflescope designs.

Weapon Sights

Unity-power optical sighting devices are frequently used on military carbines for close-quarter combat where quick target acquisition is vital but aiming errors are less critical. These sights allow an aiming reference to be seen superimposed on the target even when the eye is not precisely aligned axially or transversely with respect to the beam projected by the optical system.

One such device is the **red dot sight** that provides a large-diameter collimated beam from a battery-powered LED superimposed on a unity-power view of the target area. The sight is attached to the top of the weapon, much like a riflescope.

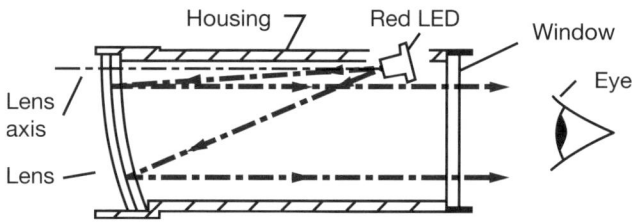

The doublet lens is designed to act as a plane-parallel window for the transmitted beam. The LED beam reflects from the interior surface of the lens and is seen as a red dot subtending a few arcminutes. Both eyes remain open so that the field of view is that of the unaided eye. If the dot can be seen, it can be used as the aiming point. Spot brightness can be adjusted for visibility against the ambient scene.

A similar device is the **holographic sight**, which has a hologram of a reticle pattern on a window sheltered within a protective housing on top of the weapon. The variable-intensity pattern appears superimposed on the transmitted unity-power view of the target when a laser powered by an internal battery illuminates the hologram.

Refracting Form

Advantages of the **refracting astronomical scope** are simplicity, ruggedness, a closed tube, good correction of aberrations, and freedom from central obscuration. Typically, it has an **achromatic** objective with a 60-mm to 100-mm aperture and focal ratio from $f/5$ to $f/15$.

> An achromat comprises two elements (cemented or air spaced) that have the same image distance (BFD) for two colors (red and blue). An apochromat has three air-spaced elements and has the same BFD for three colors (red, blue, and yellow). Secondary chromatic aberration measures the axial separation between the yellow and red/blue images in each type of lens.

Secondary chromatic aberration for an achromat typically is ~EFL/2000. This aberration limits the f/number to ~0.122 times the diameter of the objective in millimeters. For apertures larger than ~100 mm, the length of the achromatic scope is objectionable. At low magnifications, this aberration is not so distracting, so ~$f/4$ systems can be used.

The color correction of the **apochromatic** objective is better than that of an achromat. Secondary color can be reduced to zero. Focal ratios of $f/5$ can be used. This lens type is a popular choice for wide-field astrophotography at prime focus.

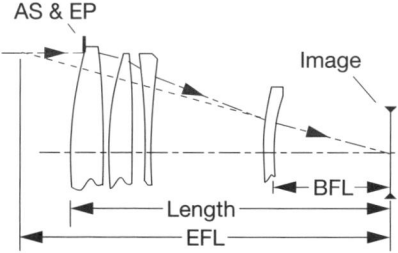

The apochromatic and **telephoto** concepts are sometimes combined in expensive binoculars to give good imagery in a short package. The last element (sometimes a doublet) must be negative and separated from the other lenses. This makes the EFL greater than the length.

Newtonian, Cassegrain, and Gregorian Forms

The **Newtonian** objective has a parabolic primary and a flat mirror at 45° as shown (below left). That mirror is held in place by a support (**spider**) attached to the entrance aperture of the scope tube (not shown). It may have one, three, or four narrow vanes crossing the scope aperture. The eyepiece is near the front of the tube.

Coma limits the RFOV of this objective. The coma should not exceed 1 arcsec, which is the usual value for atmospheric seeing. For a relative aperture of $f/5$, the usable RFOV according to this criterion is only ~0.07° in diameter. Coma-correcting attachments are available to reduce this aberration and increase the field.

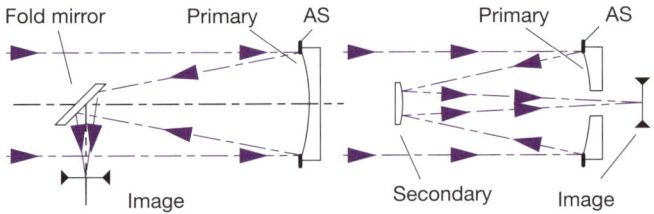

The **Cassegrain** objective (above right) has a concave parabolic primary and a convex hyperbolic secondary mounted coaxially within the tube. The secondary is supported by a spider.

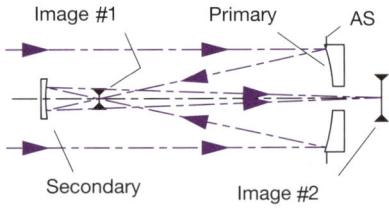

The **Gregorian** objective (below) also has a concave parabolic primary. Its secondary is a concave hyperbola that is also supported on a spider. The classical Cassegrain and Gregorian forms are rarely used in amateur astronomy scopes. They usually are used only in larger professional scopes.

Schmidt–Cassegrain and Schmidt–Gregorian Forms

The most popular form of the **Schmidt–Cassegrain** (Sch.–Cass) scope has a thin aspheric corrector plate located at or near the focus of the primary mirror. Its compactness, good optical performance, and closed tube make it a favorite of amateur astronomers. Some primaries are spherical.

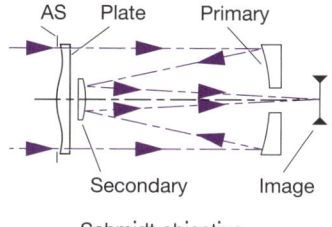

Schmidt objective

A typical scope of this type has an aperture of 200 mm and is $f/10$. System length is ~430 mm and weight is ~5.7 kg. The corrector plate reduces the off-axis spot size to about 25% of that of a true Cassegrain of the same f/number. A relatively long-focus eyepiece is necessary to obtain a wide field of view. For example, a 55-mm focal length Plössl eyepiece with a 46.0-mm diameter field stop gives an RFOV of 1.3° and an AFOV of 45.2°.

Focus of the scope is best adjusted by moving the primary mirror axially. This minimizes spherical aberration when the scope is used for terrestrial observing at close distances. Some systems eliminate diffraction and obscuration from a spider by attaching the mirror to the plate.

The Schmidt–Cassegrain is an obscured system, with a secondary obscuration typically ~33% of the primary diameter. This reduces quality and contrast of the image. The slow f/number and a curved field are drawbacks for photographic use. Commercial focal reducers and correctors for field curvature are available to improve performance in this mode of operation.

Schmidt–Gregorian (Sch.–Greg.) objectives are less common than the Sch.–Cass. The larger length of the former is a contributing factor here. Here, also, the primary may be spherical. The Zeiss spotting scope shown on page 30 is an example of the use of this form.

Maksutov–Cassegrain Form

The **Maksutov–Cassegrain** (Mak.–Cass.) scope features a meniscus lens corrector inside the focus of the spherical primary mirror. This corrector reduces the spherical aberration of the primary mirror. An advantage of the Mak.–Cass. over the Sch.–Cass.

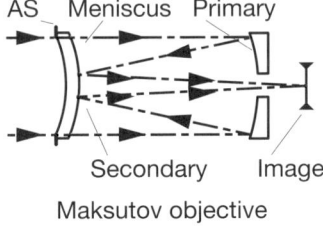

of the same f/number is higher image quality, hence a smaller image spot size. All other advantages and disadvantages are similar to those of the Sch.–Cass.

The cost of the Mak.–Cass. tends to be higher than that of the equivalent Schmidt-Cass. For example, at an aperture of 200 mm, the Mak.–Cass. currently (as of 2011) costs about four times as much as the Schmidt–Cass. The f/number of the Mak.-Cass. telescope is from 10 to 15, so the field of view is usually narrower when compared to other types of telescopes. An adjustable primary mirror is often used in Mak.–Cass. telescopes to adjust focus. The Mak.–Cass. is somewhat more stable with respect to optical alignment than the Sch.–Cass.

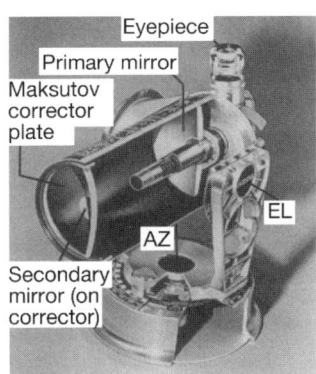

This type of scope is often used for amateur as well as professional imaging applications due to its ruggedness, compactness, and exceptional correction of aberrations. Many such scopes with apertures <100 mm provide great convenience in use and high performance in very portable packages. One design, the 89-mm aperture Questar shown here, is so popular that it has been in production for over fifty years.

Richest-Field Form

A **richest-field(tele)scope (RFT)** is defined as one that shows the maximum number of stars within the available field of view. Originally, the RFT was intended for amateurs to see a pleasing view of the Milky Way.

The size of the field of view is set by the lower limit of magnification, which is determined by the maximum size of the dark-adapted eye pupil (7 mm). There is also a limit on the telescope f/number set by the relationship between field lens diameter and eyepiece focal length. For a 70° AFOV, the maximum focal length for a 31.75-mm diameter eyepiece is 23 mm. This implies a maximum f/number of f/3.3 for a 7-mm pupil. For a 50.8-mm eyepiece, the f/number for a 70° field of view is f/5.4.

> The refractor and Newtonian forms of telescopes are best for use as RFTs because the f/numbers of all types of Cassegrains would be too slow.

Recent RFT calculations assuming a 7-mm pupil and 70° AFOV show the aperture of the optimum RFT to be ~265 mm and the number of visible stars to be ~1150, as shown here.

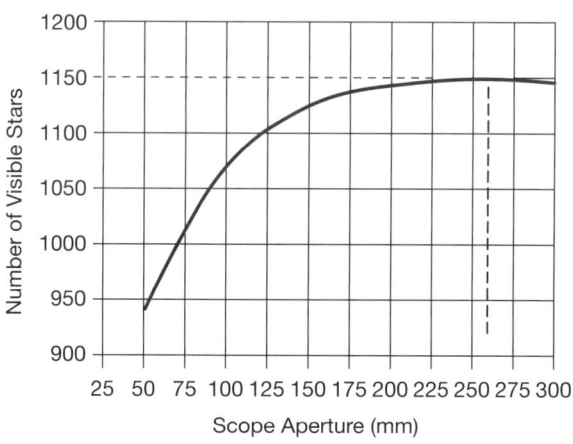

Light-Duty Mounts

Binoculars and small scopes are often mounted on photographic tripods with **heads**. The most popular head types are the **ball** and the pan/tilt. A drawback of both types is that the centers of gravity of the instruments are above both axes of motion. They tend to overturn easily if not securely clamped. For stiffness, the tripod and head should each have a load capacity at least twice as great as the weight of the instrument. See page 43 for other attributes of the tripod.

The ball head is a spherical ball-and-socket joint. It offers flexibility in rotation about elevation and azimuth axes, as well as limited roll about an axis through the ball's center and parallel to the instrument's optical axis. These motions are difficult to control independently because of crosstalk. In the ball head, all motions are locked and unlocked by twisting the handle with which the user points the instrument. Two-handed operation helps the user control unwanted motions.

Pan/tilt heads have separate elevation and azimuth rotation axes that can be independently adjusted and locked. Typically, a pan/tilt head can be more accurately controlled than a ball head. Some of these heads are fitted with a fluid damper to promote smooth motion when following moving objects. A parallelogram mechanism placed between the mount and the tripod facilitates vertical adjustment of a large binocular without disturbing the pointing of the LOS. An example is shown here.

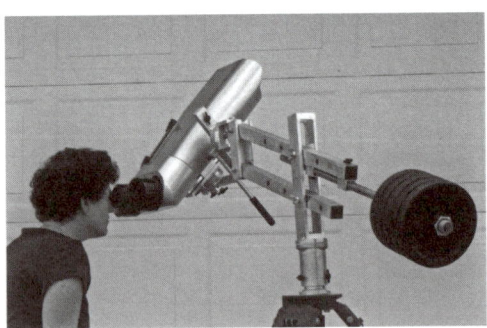

Heavy-Duty Mounts

An astronomical binocular or scope requires a sturdy support. A **heavy-duty** tripod and head provides this in portable form. Permanent mountings might be on an angled concrete pier. The basic head types are (1) **altitude over azimuth (Alt/Az)**, (2) **equatorial**, (3) **German equatorial mount** (**GEM**), and (4) **fork**. Each type has two motion axes perpendicular to each other. Most benefit from an inclined eyepiece for observation when pointing to the zenith. The instrument support arms must be long enough for the eyepiece to swing through above the tripod.

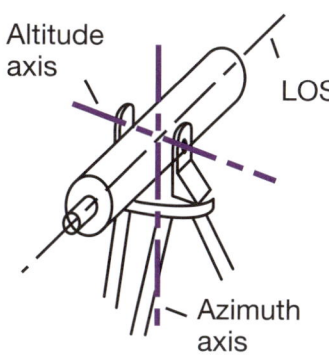

One axis of the Alt/Az head is plumbed parallel to the local vertical. Rotation about this axis provides azimuth (horizontal) motion of the scope's LOS. Rotation about the other axis swings the scope vertically. To follow a star, a zig-zag path using motions about both axes is needed. This is not easy if done manually, but it poses no significant problem for a computer-controlled **GOTO drive** in which a computer locates and tracks a designated celestial object automatically. See page 47.

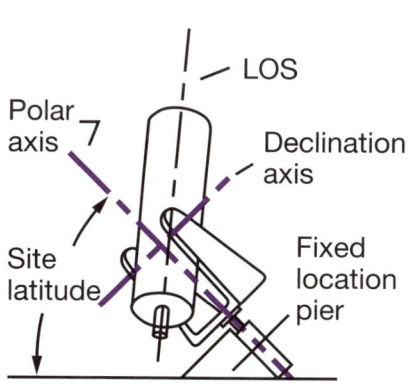

The equatorial head is a variation of the Alt/Az head in which the vertical axis (now called the **polar axis**) is tilted parallel to the Earth's rotation axis, i.e., at the site's **latitude**. The star field is stationary in the scope's FOV if this axis is driven at the Earth's diurnal rate of 23 h, 56 m per rotation.

Heavy-Duty Mounts (cont.)

The **declination** axis is perpendicular to the polar axis and allows the LOS to scan transversely to the diurnal motion.

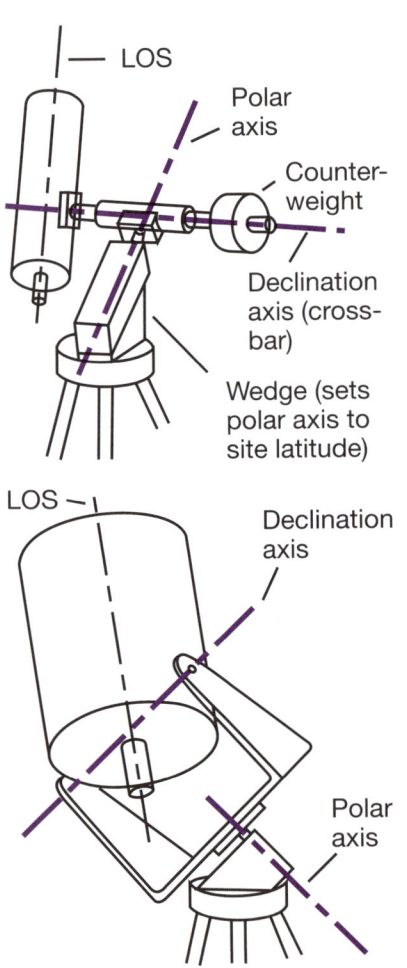

LOS

Polar axis

Counter-weight

Declination axis (cross-bar)

Wedge (sets polar axis to site latitude)

LOS

Declination axis

Polar axis

The GEM is another variation of the equatorial head. It is "T" shaped with the polar axis as the long leg and the declination axis as the crossbar. The scope is carried at one end of the crossbar. A counterweight is placed at the other end of the crossbar for balance. This head is relatively compact and stiff but heavy due to the counterweight.

The equatorial fork head is shown here. It does not need a counter-weight and is not as stiff as the GEM, due to the long overhang of the scope with respect to the polar bearing. This head is best used with short, compact scopes such as the Sch.–Cass. or the Mak.–Cass.

If used with a refractor, the fork arms would be impractically long. Some fork mounts are designed to be used as an Alt/Az or tilted back for use as an equatorial. This can sometimes be done by changing the tripod leg lengths.

Tripod Attributes

Important attributes for a tripod are load capacity, vibration response, size, weight, and cost. Its load capacity should be at least twice the instrument weight. Furthermore, the fewer the joints in the tripod, the greater the tripod's stability.

The tripod's vibration response under loads such as wind gusts (or even camera shutter operation) degrades optical performance of the supported instrument. The vibration response is determined by the **natural** (or **fundamental**) **vibration frequency** f_n of the tripod and its **damping**. Damping influences the **settling time** (time required for vibration amplitude to decay to 1% of its initial value). These parameters are determined by the tripod's structural design and materials. Hollow sections have greater stiffness-to-weight ratios than solid sections. The natural frequency f_n of a tripod is proportional to $[(E)(I)]^{0.5}$ where E is the material's **elastic modulus** and I is the cross-sectional **moment of inertia** of one leg (see table).

Material	ρ (kg/m$^3 \times 10^3$)	E (GPa)	η
Aluminum	2.71	68.9	2.5×10^{-3}
Wood	0.59	12	8.6×10^{-3}
Composite	1.78	93	2.0×10^{-3}

Settling time T_S is proportional to $1/[(f_n)(\eta)]$ where η is the material's **damping coefficient**. Generally, at comparable weight, the f_n of an aluminum tripod is superior to that of a wood tripod. Due to greater damping, the settling time of a wood tripod is shorter than that of an aluminum tripod and is intermediate for composite.

For example, at constant weight, the thickness of the walls of a hollow square aluminum tripod leg is about 5% of the width of the leg when compared to a solid wood tripod leg. The frequency ratio of the two legs is $[(E_{AL})(I_{AL})/(E_{WOOD})(I_{WOOD})]^{0.5} = 1.6$. The settling time ratio is $[(f_{AL})(\eta_{AL})]/[(f_{WOOD})(\eta_{WOOD})] = 0.47$. The f_n for the aluminum tripod is 1.6 times higher than that of a wood tripod, but its T_S is about twice as long.

More about Equatorial Mounts

This figure shows a 152-mm aperture, $f/12$ Mak.–Cass. scope mounted on a heavy-duty aluminum tripod with a German equatorial head.

The main advantage of the equatorial head, and especially the GEM version thereof, is the ease of following motions of objects in the sky. The motions of its axes correlate with celestial equatorial coordinates. This is an aid when finding objects or using a star chart. Another major advantage is that in photography there is no image rotation.

Effective use of an equatorial head requires accurate alignment of the polar axis with the axis of the Earth. In daylight, this can be done with a built-in magnetic compass and level. Other equatorial heads have a small telescope with a reticle built into the polar axis, which is hollow. This **polar alignment scope** is used at night to adjust the scope's polar axis by pointing at Polaris. Subsequent perceived motion of the stars in the field provides information on residual misalignment. This method is called "**drift alignment**."

The GEM is relatively compact and stiff. A disadvantage is that a long scope may run into a tripod leg or pier when pointed to or near the meridian. Another disadvantage is the need for the counterweight to balance the system. Use of a longer crossbar and a lighter weight to provide the proper angular moment for balance increases potential size.

More about Equatorial Mounts (cont.)

In the classic fork version of the equatorial mount, the scope is carried between the U-shaped arms of the fork so no counterweight is necessary. This economizes weight. The arms need to be long enough for the meridian to be crossed without concern about running into the tripod or other support. Another drawback is that space behind the scope is limited because the scope must be able to swing between the arms of the fork. This can be an issue when attaching a camera to the scope.

A more serious disadvantage is the relatively long length of the fork arms when used with large-aperture-diameter scopes, which lowers stiffness and increases susceptibility to vibration. The fork mount works well with large astronomical binoculars such as that shown on page 40. Because of the 45°-tilted eyepiece arrangement, access to the zenith is not compromised. It also works well with large spotting scopes in terrestrial applications.

Some fork mounts are designed so that they can be used as Alt/Az mounts, or tilted back for use as an equatorial version. For example, the 89-mm aperture, $f/14.4$ Questar scope (see photograph on page 38) has three legs that screw into appropriately oriented holes in the base to form a conventional short tripod. The scope then functions in terrestrial applications as if on an Alt/Az mount.

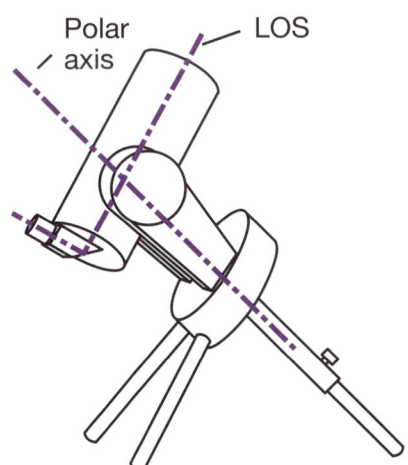

Polar axis   LOS

By switching the legs to other holes, the scope will function as if on an equatorial mount in many astronomical applications. The latter configuration is shown here schematically. The longer leg is easily adjusted to set the polar axis for the site latitude.

Dobsonian Mounts

The **Dobsonian** scope mount consists of three major parts: a rocker box, a bearing box, and a tube assembly. It was conceived as a low-cost, portable mount of Az/El (azimuth over elevation) type for Newtonian optics. Originally homemade of plywood modules, the disassembled scope can easily be transported and reassembled in the field. Later models feature tubes made of Sonotube® (cardboard form for making concrete pillars).

This sketch shows the basic configuration of a Dobsonian mount. The azimuth and elevation bearings are constrained only by gravity and contain Teflon® pads to reduce friction. No drives are provided.

Commercial variations of this design with apertures to >300 mm feature tubes with one- or two-tier open metal trusses that connect a mirror module to a forward module containing the diagonal and the eyepiece. This structure can easily be disassembled for transit and reassembled on site for viewing.

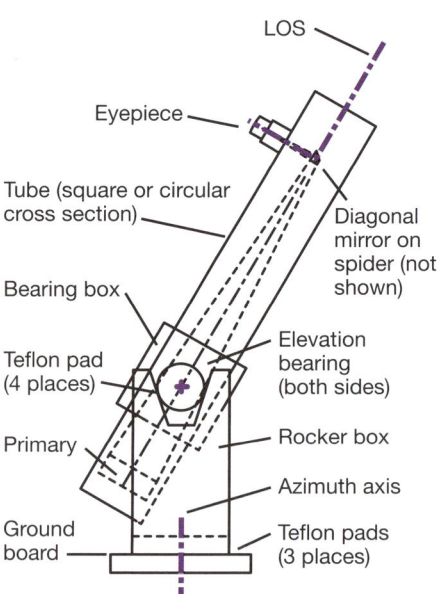

More elaborate models have electronic shaft encoders on both axes to indicate their azimuth and elevation. The user moves the scope LOS manually to the predetermined coordinates of the celestial object of interest. Thereafter, that object is manually tracked.

GOTO Drives

Finding dim celestial objects is sometimes difficult since they may be below the limiting magnitude of the eye and the FOV of the scope is small. Scopes that have **setting circles** or **encoders** to give the angular positions of the axes can use those tools to point the LOS at any object for which the equatorial coordinates and the local **sidereal time** are known.

Because portable scopes may not be precisely aligned with the Earth's axis and sidereal time is not easily determined, setting circles are not as widely used in today's amateur astronomy. With the advent of pocket calculators in the 1970s, amateur astronomers were able to quickly perform the calculations associated with the use of setting circles, including correction for known instrumental alignment errors.

In 1987, Celestron introduced the Compustar line of scopes that used a computer to calculate the position of celestial objects and then point an equatorial mount using motorized axes. This automated mount for pointing a scope is now called a **GOTO mount**.

A GOTO mount requires manual entry only of the local time and the latitude and longitude of the site. In the latest designs, the latter is provided by a **GPS**. The alignment of the scope is then calibrated by pointing it at one or two known bright stars. After calibration, a target can be found by entering its coordinates into the control system or, more commonly, by selecting it from the control system database. The mount then slews automatically to that position. This saves time, helps beginners, and often enables experienced observers to find very faint objects.

The database of a typical GOTO mount may contain information regarding 29,000 to >147,000 targets. Slewing speeds range from 5° to 8°/sec while pointing accuracy may be 1 to 5 arcmin, depending on scope alignment accuracy and system mechanical parameters.

Stereoscopic Vision through a Binocular

When two objects located at differing distances from the observer are seen through a binocular, the disparities between the retinal images are increased and depth perception is improved over that of the unaided eyes. These changes result from the magnification introduced and by the increase in baseline.

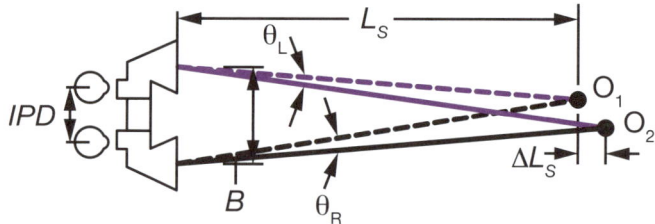

The minimum detectable separation ΔL_s at distance L_s and the maximum L_s, both in meters, for **aided stereo vision** to occur are given approximately by:

$$\Delta L = 1000 L_S^2 \Delta\theta / [(MN)(IPD)]$$
$$L_{MAX} = (N)(IPD) / [1000(\Delta\theta / M)]$$

where $N = B/IPD$, $\Delta\theta$ is the unaided eye stereo acuity in radians [usually ~10 to ~30 arcsec (~ 4.85×10^{-5} to ~1.45×10^{-4} rad)], and M is the magnification. $L_s, \Delta L_s$, and L_{MAX} are in meters.

> The example from page 13 is repeated here for a binocular having $N = 1.77$ and $M = 7\times$. The $IPD = 65$ mm, $\Delta\theta = 30$ arcsec (1.45×10^{-4} rad), and $L_s = 800, 1600, 3200$, and 6400 m. Then, $L_{MAX} = $ ~5554 m and $\Delta L_s = $ ~115, ~461, ~1843, and ~7373 m respectively. The advantage of using the binocular is apparent.

As in the unaided case, cues other than stereo vision (relative size, perspective, parallax, atmospheric effects, etc.) significantly influence depth perception.

Resolving Power with Optics

In order to describe how the combination of an afocal optical system (binocular or scope) with the eye is able to resolve details in a distant object, we introduce the concept of **modulation transfer function** (**MTF**). This parameter takes into account the fact that any real object has coarse and fine details—all of which provide information to the brain that interprets that image.

Consider the simple three-bar object shown in view A below. It has high contrast, i.e., black lines and white spaces (line pairs). Scanning across this object, as indicated by the arrow, we see a square-wave luminance variation (view B). Because of aberrations, and perhaps diffraction, the eye and the optical system blur the pattern while forming its image. The redistributed energy is represented by a sine curve that has maximum and minimum values slightly smaller and larger than the corresponding values in the object. The illuminance in the image is said to be modulated with a modulation factor expressed as (max – min)/(max + min). View C shows how the modulation factor decreases as the structure of the object gets finer, i.e., has a higher **spatial frequency**.

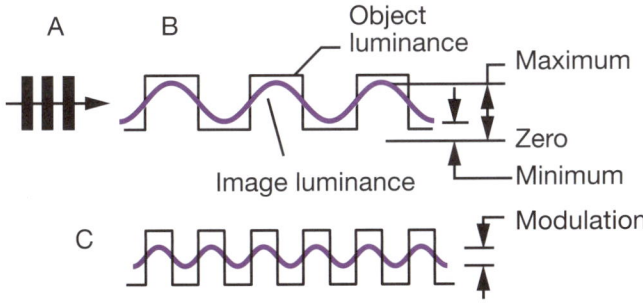

MTF is defined as image modulation/object modulation and is usually plotted against spatial frequency in cycles/mm or cycles/mrad. Capability to calculate MTF is provided in lens design programs.

Resolving Power with Optics (cont.)

The first step for applying this technique to visual systems is to determine the MTF of the objective and erecting prism subsystem (if used) at its focal plane. The design for the eyepiece is then combined with a math model of the eye to find the modulation required at that same plane to resolve the frequencies in the image. The latter information is known as an **aerial image modulation** (**AIM**) curve. The intersections of the MTF and AIM curves denote the limiting resolvable frequency.

This graph shows the MTF for a 1200-mm EFL f/10 achromatic scope objective (black curve) combined with AIM curves for an eye model plus a series of four eyepieces of similar design, but differing EFLs (colored curves). The combinations yield four astronomical scopes of 120-mm apertures and magnifications ranging from 43× to 150×. The angular resolution in object space varies from 2.7 arcsec/lp (low) for the 28-mm eyepiece to 1.5 arcsec/lp (high) for the 8-mm eyepiece. This shows that magnification improves the system's resolution.

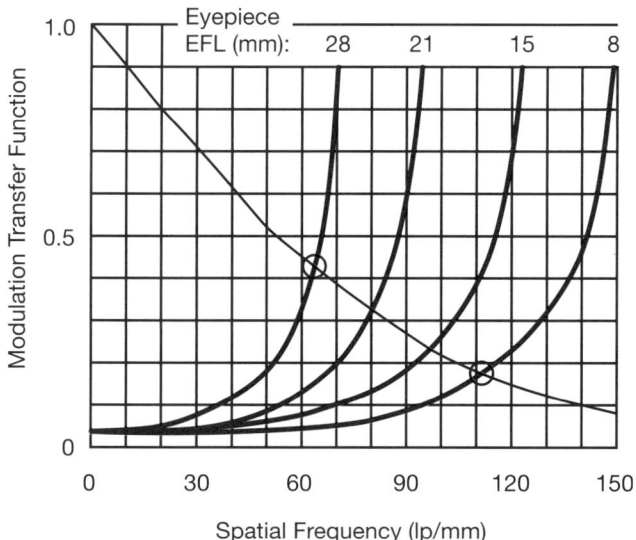

Binocular/Scope Efficiency

Binoculars and scopes extend the ranges R_{OPT} at which an object of given size and contrast with respect to its background can be detected with the optic as compared to the detection range R_{EYE} for that object using the unaided eye. Efficiency E is defined as the ratio $E = R_{OPT}/R_{EYE}$. The instrument also increases the amount of detail in the given object that can be seen by the user at a given range. This increase may be expressed in terms of the smallest object detail resolved, i.e., the **visual acuity** or **limiting resolution** V_{OPT} with the optic and with the unaided eye V_{EYE}, i.e., $E = V_{OPT}/V_{EYE}$.

Research has shown that E is determined primarily by the instrument's EP diameter D_{EP}, its magnification M, its light transmission T, and the **scene luminance** L. The general equation for E is:

$$E = M^{1-X} D_{EP}^{X} T^{Z}$$

Three luminance conditions are of greatest importance: daylight ($L > 0.03$ cd/m^2), twilight (0.001 cd/m^2 $< L <$ 0.03 cd/m^2), and night ($L < 0.001$ cd/m^2). Applicable equations for E are:

$$E_{Daylight} = MT^{0.25}$$
$$E_{Twilight} = (MD_{EP})^{0.5} T^{0.33}$$
$$E_{Night} = (MD_{EP}/D_{Eye})^{0.5} T^{0.5}$$

These expressions allow us to compare how various instruments should perform under specific conditions.

The user usually does not know the transmission T of a particular binocular or scope because manufacturers do not often specify it. It can be estimated using techniques described on page 76 and 77. As shown there, T for a typical 7×50 binocular with cemented BaK4 Porro prisms is ~91%. Similar calculations show T for a 10×50 binocular with cemented BaK4 Abbe–König prisms to be 89%.

Binocular/Scope Efficiency (cont.)

Two binoculars are compared here for twilight use. Unit 1 is a 7 × 50 with transmission of 91% and Unit 2 is a 10 × 50 with transmission of 89%. Then, $E_2/E_1 = [(10)(50)]^{0.5}(0.91)^{0.33}/[(7)(50)]^{0.5}(0.89)^{0.33} = 21.52/18.14 = 1.19$. The 10× binocular would be expected to allow the user to detect a given target at a range ~20% greater than the same user could detect the same target with the 7× binocular.

The night glass is used at very low luminance levels. The resolution of the eye is then degraded by a factor of 10 or more from its daylight capability (see graph on page 10).

Magnification increases the apparent size of the object, making it easier to see. For observation at night, the XP diameter of the instrument should equal that of the user's dark-adapted eye pupil. This diameter is commonly assumed to be 7 mm even though it was shown on page 8 that it depends on age.

The US Army's WWII night glass, the M17 7 × 50 binocular, is shown above. It is a militarized version of a nonmilitary Bausch and Lomb design. Similar models were used by the US Navy and Marines.

Note that, in the above calculations, certain inevitable factors that influence efficiency have been neglected. These include contrast reduction in the atmosphere, optical image quality loss due to residual aberrations, stray light effects, and the effects of vibration, including muscular tremble.

Handheld-Binocular Efficiency

Muscular tremble or **shake** reduces the efficiency of a binocular or scope when handheld in comparison to that achieved when the instrument is solidly supported on a tripod or another mount. The supported case applies approximately when the optics are stabilized by means such as those described on pages 85 and 86. Only angular motion is important here.

> Angular motions of a handheld instrument are sinusoidal, with a frequency of about 9 Hz and magnitudes of about 0.25°, peak to peak. Size and weight of the instrument do not significantly affect these values, but they affect ease of use and fatigue.

The daylight efficiency of a handheld instrument of magnification M is related to its daylight efficiency when supported as

$$E_{Handheld} = E_{Supported}/(1 + 0.05M).$$

Combining this equation with that for daylight efficiency from page 51 gives:

$$E_{Handheld\ Daylight} = MT^{0.25}/(1 + 0.05M).$$

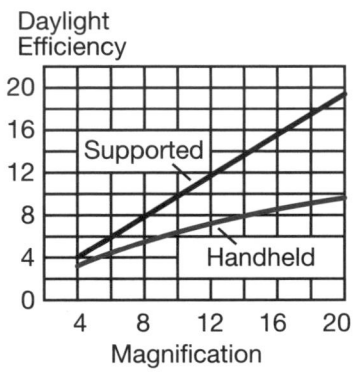

Daylight Efficiency

The calculated daylight efficiencies for supported and handheld instruments with magnifications of 4× to 20× and transmissions of 91% are plotted here.

Distortion Effects

Distortion does not affect the quality of an image, but displaces it from its normal position or changes its shape.

There are two types of distortion in optical instruments: angular and rectilinear. These cannot be corrected simultaneously. **Rectilinear distortion** changes the image geometry: straight lines become curved. The two types of rectilinear distortion are barrel and pincushion.

To correct this aberration, the ratio $\tan\alpha/\tan\beta$ must be constant and equal to M. Terrestrial binoculars and scopes are preferably corrected for this type of distortion.

A disquieting dynamic visual illusion seen through many binoculars and scopes is called the "**globe effect**" or "**rolling ball effect**," which occurs only in an instrument corrected for rectilinear distortion. As the instrument LOS is scanned, targets appear to be moving on the surface of a convex sphere. This illusion arises from the change in velocity and size of an object as it moves from the center to the edge of the FOV. The increase Δv in relative image velocity from the axis to the edge of the field is expressed as $\Delta v = 1/[\cos^2\beta + (M^2\sin^2\beta)]$. Pincushion distortion is sometimes deliberately introduced to reduce this illusion.

Astronomical instruments are preferably corrected for **angular distortion**. Then, the shapes of images such as those of star clusters are not distorted at the FOV edge.

Angular distortion changes the M of the image with increasing image height. For zero angular distortion, the ratio α/β must be constant over the field.

Residual distortion can be calibrated out of digital images by computer processing if necessary.

Limiting Magnitude of a Binocular or Scope

The **apparent visual magnitude** M_V of a celestial object is a measure of its luminance, at a wavelength of 550 nm, as observed by the eye from Earth. A difference of one magnitude is defined as a change in luminance by the fifth root of 100 (2.512). Objects with lower magnitudes (which can be negative) have higher luminances, so appear brighter than ones with numerically higher magnitudes.

The luminances of extended objects in the sky, such as galaxies and star clusters, are given as integrated magnitudes with all of the light concentrated into a single point. Units are magnitude/(arcsec)^2.

Object	M_V
Sun	-26.5
Full Moon	-12.2
Venus	-4.3
Sirius	-1.6
Dimmest object visible in suburban setting	3.0
Dimmest object visible in rural setting	5.0
Dimmest Messier object (M97)	11.0

The luminance of the dimmest star visible through a binocular or scope is called the **limiting magnitude** M_L of that instrument and is approximated as

$$M_L = M_E + 5\log[(TD_{EP})/D_{EYE}],$$

where M_E is the visual magnitude of the dimmest star visible with the unaided eye, T is the transmission of the instrument, D_{EP} is the diameter of the instrument's entrance pupil, and D_{EYE} is the pupil diameter of the eye pupil. This relationship assumes binocular vision.

It has been suggested that binocular vision should increase object luminance by $\sqrt{2}$ or ~0.38 magnitude, but extensive testing during WWII showed this increase to be only ~10% or ~0.1 magnitude.

Limiting Magnitude of a Binocular or Scope (cont.)

The optical instrument does not change the apparent contrast ratio of an extended object such as a nebula or galaxy, and therefore does not make such an object appear brighter. When the scene luminance decreases, the resolution of the eye also decreases (see page 10). A scope or binocular improves visibility of dim extended objects in the sky by making them appear larger. As a rough rule of thumb, the apparent angular size of the object as seen through the binocular or scope should be ~1° for detection.

Experience indicates that the limiting magnitude predicted by different observers will vary by ~1.5 magnitude. The following equation has been suggested for 50% probability of detection of faint stars through a scope of aperture D_{EP} (in mm), assuming optimum conditions (experienced observer, 70% instrument transmission, XP and eye pupil diameters of 7.5 mm, and an unaided eye limiting magnitude of 8.5):

$$M_L = 3.7 + 5\log D_{EP}.$$

This graph shows M_L for 50% probability of detection:

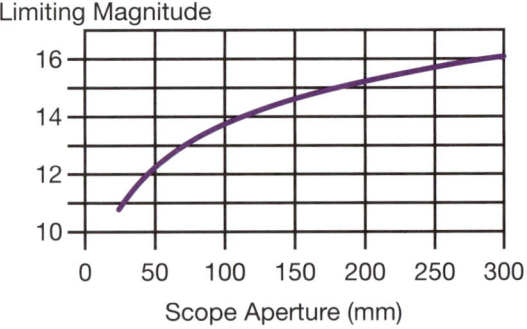

Diffraction Effects

The resolving power of a perfect optical system is determined by **diffraction** that results from the wave nature of light. An infinitely distant point source will image as a central peak surrounded by dark and faint bright rings, called the **Airy disk pattern**. Its intensity distribution and a theoretical image of a point source are shown here.

Relative Intensity

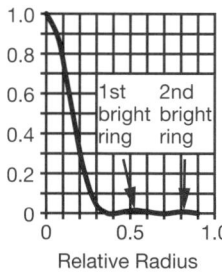

Relative Radius

If the **Rayleigh criterion** is applied to resolution of double stars of equal luminance, their patterns overlap so that the peak of one lies at the first minimum of the other. The angular separation θ_R in arcseconds is then $\theta_R = 138/D_{EP}$, where D_{EP} is the instrument's entrance pupil diameter in mm. The empirical **Dawes criterion** for these double stars $\theta_D = 116/D_{EP}$ supports this theory. Either of these criteria can also be applied approximately to observed detail in terrestrial objects.

The minimum resolution of the eye is about one arcminute. The minimum magnification necessary to resolve images at the diffraction limit is dependent on D_{EP} and is ~0.43× per mm of aperture. This would be ~22× for a 50-mm objective. Higher magnification, such as 2× per mm, makes observing details easier for the eye. The excess of magnification beyond 0.43× per mm of aperture is called **empty magnification** and does not improve resolution.

Aberrations of the optical system reduce the height of the Airy disk central maximum. The **Strehl ratio** is the ratio of the actual height to the theoretical height. A ratio of 80% represents the **Marechal criterion**.

Obscuration Effects

Diffraction (see page 57) applies to unobscured apertures. Cassegrain and Newtonian scopes have circular central **obscurations**. They decrease **transmission** T, reduce the Strehl ratio S, and lower MTF in the images. We define obscuration ε as the ratio of the obscuration diameter D_{OBS} to the entrance pupil diameter D_{EP}. It typically ranges from 0.20 to 0.33, the values of which correspond to S of 0.92 (excellent image quality) and S of 0.80, (the Marechal criterion), with central obscuration: $T_{OBS} = 1 - \varepsilon^2$ and $S_{OBS} = (1 - \varepsilon^2)^2$.

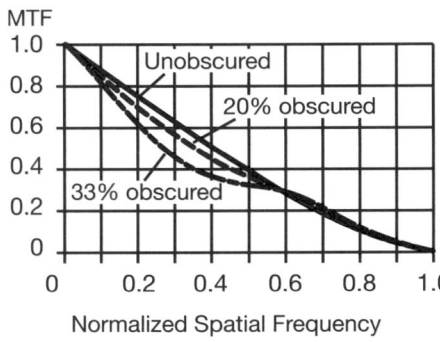

Normalized Spatial Frequency

MTF is plotted here for three values of obscuration: $\varepsilon = 0$, (unobscured), $\varepsilon = 0.20$ (typical of a Newtonian scope), and $\varepsilon = 0.33$ (typical of a Schmidt– or a Maksutov–Cassegrain scope). The normalized spatial frequency is $v_n = v\lambda/D_{EP}$. Here v and v_n are in cycles/mm.

As a rule of thumb, MTF of an obscured system at low-to-mid spatial frequencies is equivalent to that of an unobscured system with aperture diameter equal to the diameter of the obscured system minus the diameter of the obscuration.

The spider supporting a secondary or folding flat mirror also obscures a small fraction of the aperture. Typically, one, three, or four vanes are symmetrically arranged about the axis to provide progressively stiffer support for the mirror. Each vane produces two **diffraction spikes** in star images. These spikes are always oriented perpendicular to the vane and result from diffraction by the straight edges. Curved vanes can reduce the spikes.

Atmospheric Scatter Effects

Light scatter by atmospheric haze reduces contrast and degrades resolution of distant objects seen vertically or horizontally through a scope or binocular. These effects depend on the **visual range** R_V, which is defined as the distance at which the unaided eye can detect an extended dark object having 2% contrast against the horizon sky. In clear daylight, R_V is >10 km. In light haze, it is 5 to 10 km and, in moderate haze, it is 2 to 5 km.

Ricco's Law states that the product of target angular size and its contrast is constant at the detection limit. Additional magnification compensates for contrast loss.

Detection of a target at distance L through a binocular or scope requires a minimum magnification of $M = (L/r)(e^x)$ where e is the Napierian logarithm base, r is the distance at which that same target can be detected with the unaided eye, and $x = 1.956(L-r)/R_V$.

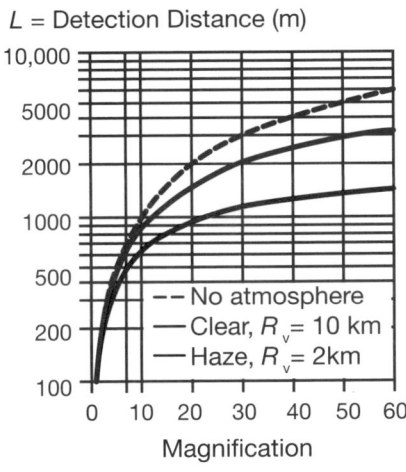

L = Detection Distance (m)

This graph shows the maximum values of L as functions of M and R_V for three atmospheric conditions when the same target can be detected at 100 m with the unaided eye. Note that a given change in M has less effect on L when the target is at a long distance.

For example, if R_V is 10 km (clear day) and $r = 100$ m, $L = {\sim}2500$ km at 40× and ${\sim}2900$ km at 50× (~16% change). For R_V of 2 km (moderate haze), $L = {\sim}1270$ m at 40× and ${\sim}1400$ m at 50× (~10% change).

Atmospheric Seeing Effects (Elevated Path)

Changes in the index of refraction of air between the observer and the target distort the transmitted wavefront and reduce image quality. This is called "**seeing**." There are three types of seeing: **scintillation** (variation in luminance), **image motion**, and **image blur**. All change rapidly with time. They are produced by local turbulence effects (due to temperature changes of the telescope and wind at the observation site), boundary layer effects in the atmosphere above the site (influenced by diurnal heating), and upper-atmospheric effects (from high-velocity winds at altitudes >10 km).

Because of seeing, the angular size of a star viewed through a perfect scope appears larger than it would from diffraction alone. Its size is then called the **seeing disc**.

The **Fried parameter** r_0 is the **coherence length** of the atmosphere. It is related to the size of the seeing disc d_S in radians at the zenith by the parameter $r_0 = 0.98\lambda/d_S$, where λ is the wavelength in mm. Scopes with apertures smaller than r_0 are diffraction and aberration limited. The Fried parameter is weakly dependent on wavelength (as $\lambda^{6/5}$) and on the angular distance z from the zenith [as $(\sec^2 z)^{-3/5}$]. During the day, the average value of r_0 is 20 to 40 mm. At night, the average value is ~100 mm. Values of $r_0 \approx 200$ mm are achieved at the best observatory sites.

The ratio D_{EP}/r_0 determines the effect of atmospheric seeing on the image. When $D_{EP}/r_0 < 3.7$, atmospheric seeing primarily produces image motion. For $D_{EP}/r_0 > 3.7$, the effect is image blur. In the latter case, the resolution is independent of telescope aperture size at ~$1.273\lambda/r_0$. Optimum resolution results at $D_{EP}/r_0 \approx 3$.

During the night, a well-adjusted amateur telescope with aperture smaller than $D_{EP} = 3r_0 \approx (3)(100 \text{ mm}) \approx 300$ mm may be able to provide diffraction-limited resolution for observing celestial objects.

Atmospheric Seeing (Horizontal Path)

When imaging along a horizontal path of length L, the Fried parameter r_0 varies with distance and is combined with another atmospheric turbulence parameter, the **index of refraction structure parameter** C_n^2 (expressed in units of $m^{-2/3}$), by the relationship $r_0 = 0.185(\lambda^2/LC_n^2)^{3/5}$. Typical values of C_n^2 range from $10^{-13}\,m^{-2/3}$ (medium effect) at the middle of the day to $10^{-14}\,m^{-2/3}$ (weak effect) at twilight.

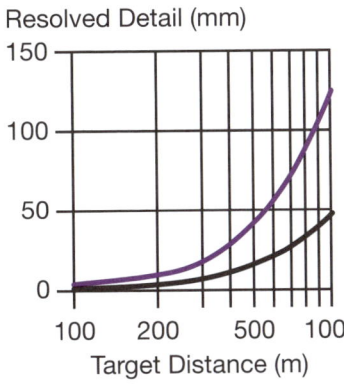

Resolved Detail (mm)

Target Distance (m)

This graph plots resolved detail (in mm) at the target distance L (in m) on a horizontal path for medium air turbulence (colored curve) and weak turbulence (black curve). As expected, increased turbulence reduces the amount of detail that can be resolved at any distance.

The **Greenwood Frequency** is correlated with the image motion in the focal plane of the viewing instrument produced by atmospheric seeing. It typically is 20 to 100 Hz. There is significant seeing-induced image motion of the image that reduces resolution during the ~50 msec integration time of the human eye.

Atmospheric turbulence severely limits the resolution of spotting scopes and binoculars at long distances. This effect is independent of the optical quality of the instrument. Since r_0 is about 20 mm on horizontal paths, the largest effective instrument aperture for maximum resolution is $D_{EP} \approx 3r_0 \approx 60$ mm.

Focusing for Different Target Locations

When observing objects at different distances through scopes or binoculars, we find that we can see objects most sharply at only one specific distance at a time. Objects nearer or farther away appear more or less blurred because their images are located at different distances from the objective. If the focus of the objective or the eyepiece is adjusted axially, the image of the selected object can be brought into focus.

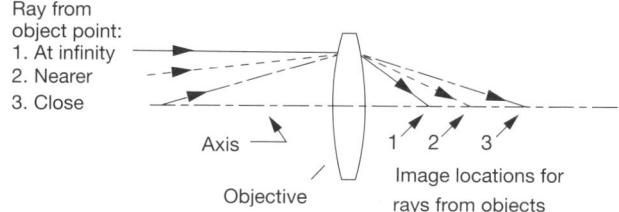

In many consumer binoculars, both eyepieces can be moved simultaneously because they are connected to the ends of a bridge that is supported at its center by a shaft built into the hinge. Turning a focus ring on the shaft moves the bridge on a thread inside the hinge. This design, called **center focus (CF)**, is convenient because it allows focus to be quickly adjusted using one finger.

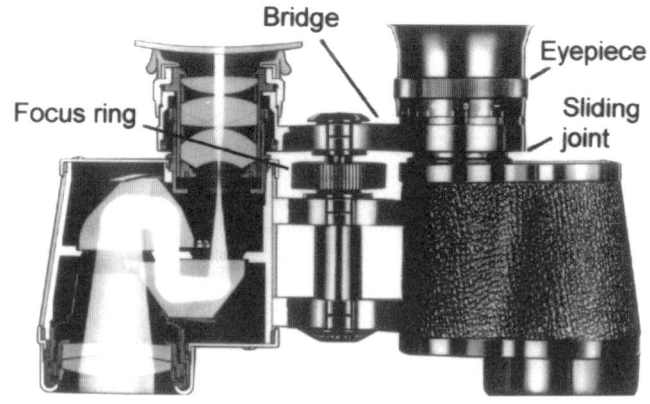

Focusing for Different Target Locations (cont.)

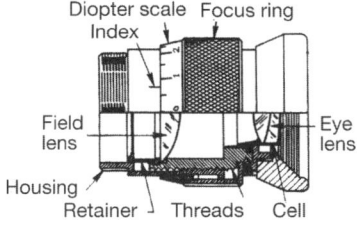

Diopter scale   Focus ring
Index
Field lens
Housing
Retainer   Threads   Cell
Eye lens

In many other binoculars, notably those intended for military, marine, or astronomical uses, each eyepiece is moved on its own internal thread to focus on the target of interest. Each eyepiece then has its own focus ring. This approach, called **individual focus**, has two advantages over center focus designs: individual eyepieces are sturdier, and the moving parts can be sealed more effectively. The example shown here is not sealed.

With **internal focus** (**IF**), each objective has one element that can be moved axially to compensate for the image shifts associated with different target distances. The "in focus" images are brought to the focal planes of the eyepieces, which remain stationary. The outermost lenses can be sealed statically. One design, shown here, moves left and right shuttles simultaneously to slide elements in both objectives. The diopter ring biases one scope's focus to correct accommodation errors. Another mechanism for accomplishing this function is described on page 99.

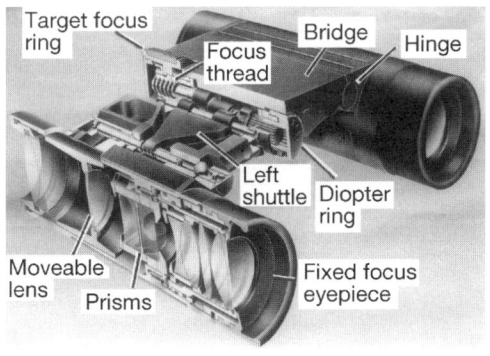

Target focus ring
Focus thread
Bridge   Hinge
Left shuttle   Diopter ring
Moveable lens   Prisms
Fixed focus eyepiece

The Diopter Adjustment

When a properly adjusted scope or binocular is placed in front of a normal eye and used to observe an on-axis point object at infinite distance, the beam entering the eye is collimated, i.e., all of its rays are parallel.

Many eyes are not normal. A **myopic** eye cannot focus properly on an infinitely distant target because its focal length is too short or the eyeball is too long. The opposite conditions prevail in a **hyperopic** eye. In order for a distant object seen through an instrument to be in focus to either of these defective eyes, the beam entering the eye must be slightly divergent for a myopic eye and slightly convergent for a hyperopic eye. Departures from the collimated condition are measured in diopters and are positive or negative.

The focusing capabilities of the user's eyes are usually different so require different amounts of compensation to bring the image into focus through a binocular. Note that the **diopter adjustment** will not correct the user's astigmatism; contacts or eyeglasses are needed for this.

Focusing the instrument with its individual diopter adjustments can compensate simultaneously for both object distance errors and the eye's refractive errors.

To compensate for eye defects, the eyepieces of most binoculars and scopes with magnifications greater than ~3× are designed to be moved axially until the image of a distant object appears to be focused. This adjustment is typically done by rotating a focus ring on the eyepiece. The movement Δ of the eyepiece, in millimeters, for 1 diopter focus change is given by

$$\Delta = (f_{EP})^2 / 1000$$

where f_{EP} is the eyepiece EFL in millimeters. An adjustment range of at least +3 to −4 diopters is needed for the instrument to be used by the adult population.

Erecting Prisms

The most popular prism designs used in binoculars and scopes are sketched here. The basic dimension for all designs is the entrance face width A. For preliminary designs, the beam

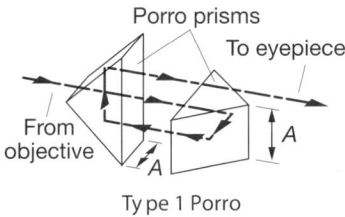

Porro prisms
To eyepiece
From objective
A
A
Type 1 Porro

is assumed to be collimated. Many systems of interest have converging beam envelopes, while wide-angle systems may have diverging beam envelopes. The prism can be customized to reflect this detail as part of system design optimization.

The axial path length is t while the axis offset is O. Glasses are BK7 ($n = 1.517$, $\rho = 2.51$ g/cm^3) or BaK4 ($n = 1.569$, $\rho = 3.10$ g/cm^3). Internal transmission $\tau = 0.998$/cm for both types. W is weight neglecting bevels and similar details.

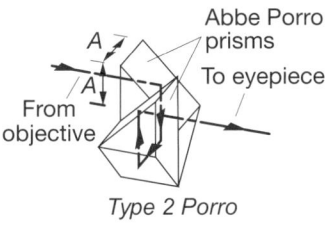

Abbe Porro prisms
A
A
To eyepiece
From objective
Type 2 Porro

The Type 1 Porro erecting system can be air spaced or cemented. $O = 1.41A$, $t = 4A$, $W = 2.57\rho A^3$, axial length = $2A$ + air space.

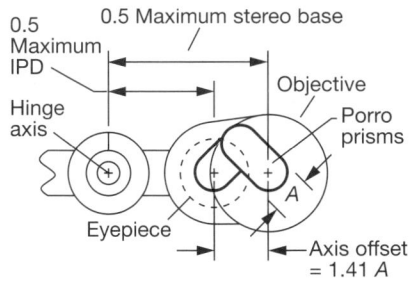

0.5 Maximum IPD
0.5 Maximum stereo base
Hinge axis
Objective
Porro prisms
Eyepiece
A
Axis offset = 1.41 A

Ernst Abbe modified the prism into the Type 2 Porro, which functions like the Type 1 and has the same basic dimensions.

In a binocular, either prism assembly type can be rotated about the objective axis to maximize axis offset as shown here. The prisms shown here have rounded ends to reduce weight.

Erecting Prisms (cont.)

Two in-line prism types are common in binoculars and scopes. First is the classic **Abbe–König** roof prism, shown below in symmetrical form. An asymmetrical form with the exit face smaller than the entrance face is used in some modern binoculars (such as the Zeiss Victory 8×56 model).

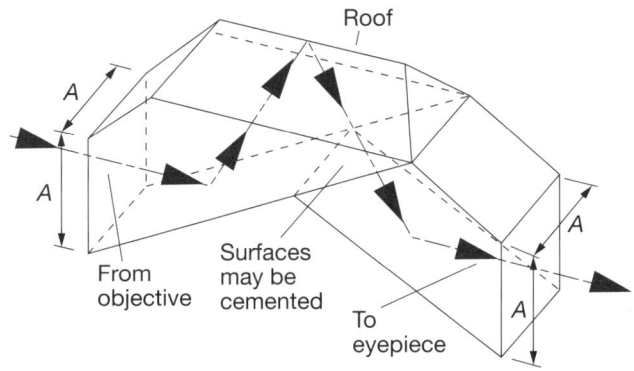

Here, $O = 0$, $t = 5.2A$, $W = 3.72\rho A^3$, and axial length $= 3.46A$.

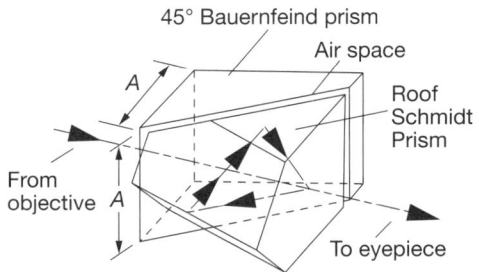

The second prism is the **Roof–Pechan**. Here, $O = 0$, $t = 4.62A$, $W = 1.80\rho A^3$, and axial length $= 1.21A$. The prisms are mechanically held with a thin air-space between to obtain two **total internal reflection (TIR)** reflections. One face has a reflecting coating.

TIR occurs when the angle of incidence of a ray passing from a high-index medium to one with lower index (such as air) exceeds the critical angle of incidence I_C, where $\sin I_C = n_2/n_1$. Beam intensity is not reduced on reflection.

Prism Refractive-Index Effects

For maximum light transmission through a binocular or scope with Porro prism erectors, the reflection from each prism surface should be TIR. The reflectivities of the four surfaces would then be 100%.

For this to occur, the incident angles with which rays in a beam from the objective strike each reflecting surface (tilted at 45° to the axis) must all exceed the **critical angle** $I_C = \sin^{-1}(1/n)$, where n is the **refractive index** of the prism glass.

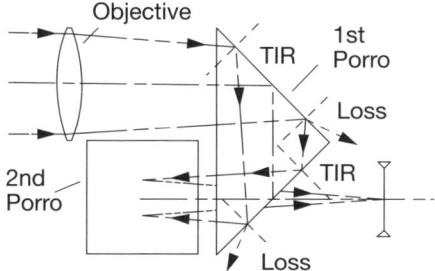

In this figure, a beam enters the objective parallel to the axis and exits converging toward the image. Its relative aperture (f/number) is a measure of the beam's cone angle. When incident on the first tilted surface, the upper ray will experience TIR; however, if the index is too small, the lower ray may not. Part of the beam is lost at each reflection. A similar occurrence takes place in the second Porro for rays orthogonal to the plane of the figure. The pertinent relationship for the f/number of a beam that always totally reflects is

$$f/number = 2\tan\{n[45° - \sin^{-1}(1/n)]\}^{-1}$$

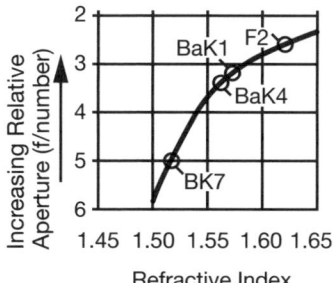

This graph shows f/number plotted versus index for four glass types. BK7 with $n = 1.517$ represents that used in old designs and lower-cost instruments today. The beam must be ~ f/5 or slower to ensure TIR.

Prism Refractive-Index Effects (cont.)

The glass of choice for newer designs is BaK4 glass with $n = 1.569$. Designs with this glass can have faster objectives up to $\sim f/3.2$. "Made in USA" versions of today's BaK1 ($n = 1.573$) and F2 ($n = 1.620$) glasses produced TIR and fully illuminated round XPs in the US Army's 7×50 M17 ($f/3.85$) and 7×50 M19 ($f/3.05$) binoculars in the 1940s, 1950s, and 1960s, respectively.

Note that, contrary to some published statements, all of the glasses named here are of equivalent high quality and light transmission if reputable manufacturers have made them. The only significant difference is refractive index. Lower-index glasses and glasses of inferior quality are found only in low-cost instruments.

One can determine if a Porro prism instrument has glass with high enough refractive index for TIR by holding it ~30 cm in front of the eye while pointing it toward an illuminated surface or the sky (not the sun). We note whether the pupil is round and fully illuminated or vignetted to a square shape. The sides of the square are parallel to the long sides of the four prism hypotenuse faces. The squared-off pupil in one binocular and the circular pupil in another binocular are illustrated here.

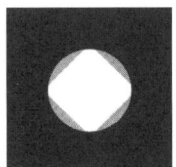 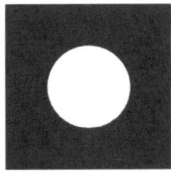

Roof prisms do not have this "vignetted XP" problem. The incident angles on the roof surfaces are large enough to produce TIR in prisms made of any of the above-named glasses. If those roof surfaces are bare, **image doubling** may be detected in the direction perpendicular to the 90° dihedral roof edge, even if the 90° dihedral angles are perfect. This doubling is corrected by applying thin film coatings to both roof surfaces. These **phase coatings** modify the physical optics behavior of the reflected light but increase cost, so they are used in more expensive models.

Lens Erecting Systems

The figure on page 33 shows the internal components of a typical riflescope. Two lenses are located between the image formed by the objective and the object plane of the eyepiece. This portion of the optical system is depicted below with important light rays designated.

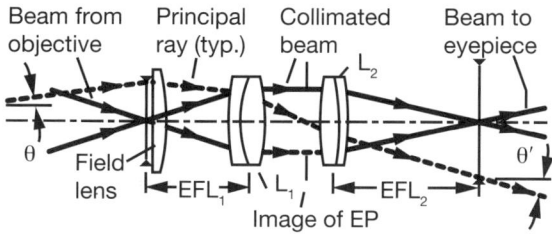

Note that a principal or chief ray (dashed line) enters moving upward at an angle θ and exits moving downward at an angle θ'. This shows that the image is turned over. In the airspace between the lenses, an image is formed of the scope's EP, which is located at the objective. All principal rays at various semi-field angles cross the axis at that location.

If lenses L_1 and L_2 have unequal EFLs and are positioned so that a collimated beam is formed between the lenses, the erecting subsystem has nonunity magnification of $M_E = EFL_2/EFL_1$. This is the case in the above figure. Equal erecting-lens EFLs would produce unity magnification. In both cases, the image is inverted.

The lens near the first image is called a **field lens**. It redirects the off-axis beam toward the axis so that the apertures of the erecting lenses can be minimized. The field lens also plays an important role in controlling aberrations by optimizing where the beams from various zones of the object intercept subsequent elements.

Erecting lenses can also form a variable magnification (zoom) system. Two lenses need to be moved synchronously along the axis to accomplish this. Such systems are discussed on page 83.

Eyepiece Configurations

Although used in some lower-cost instruments, early simple eyepieces, such as the **Huygenian** and **Ramsden**, do not provide enough variables (radii, thicknesses, refractive indices, etc.) to meet today's demands for a high-quality image in a binocular or scope. Hence, they are not discussed here. Six higher-performance eyepiece configurations are provided below (not to scale). Dimensions are relative to eyepiece focal length f_E and are approximate. Many modern eyepieces are derived from these designs.

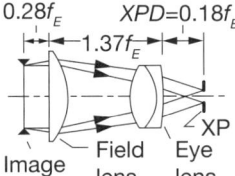

The **Kellner** design covers a 45° AFOV. It has a singlet field lens and a cemented doublet eye lens and is often used in lower-cost instruments. Its XPD is quite short.

The **orthoscopic** eyepiece (below left) has a plano-convex eye lens and a cemented triplet field lens. The latter lens may be symmetrical to reduce cost. Generally, good imagery is produced over a 45° AFOV. Distortion is well corrected.

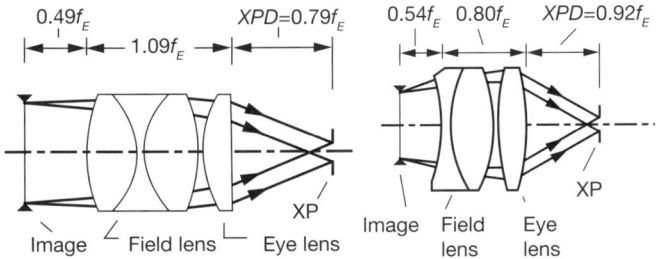

The **RKE** design (above right) is a reversed, shortened, and significantly improved version of the Kellner. It is similar in appearance to the WWI König design. Comprising a doublet field lens and a singlet eye lens, it covers a 45° AFOV with very good aberration correction and has a large XPD. The eyepiece of the riflescope shown on page 33 is of this general form.

Eyepiece Configurations (cont.)

The **Plössl**, or "symmetrical," eyepiece design (below left) has similar cemented doublets with their crown elements adjacent. It covers a 50° total apparent field.

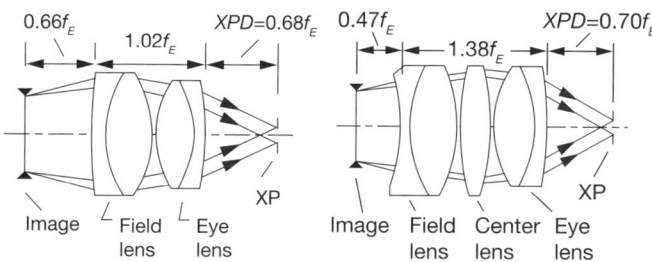

The **Erfle** design (above right) covers a ~65° AFOV. Aberrations are well corrected except distortion, which is moderate. High-index glasses, some with low chromatic dispersions (called ED glasses), give AFOVs reaching 70°.

In 1873, **Smyth** added a negative lens near and in front of the focal plane of a photographic lens to reduce field curvature and increase FOV. During WWII, eyepieces with total AFOV as large as 110° were developed with Smyth field lenses for German military scopes. Nagler significantly improved this eyepiece design in the early 1980s. His work formed the basis for a series of currently available extremely wide-angle eyepieces. A **Nagler eyepiece** with $2\beta = 78°$ is sketched here.

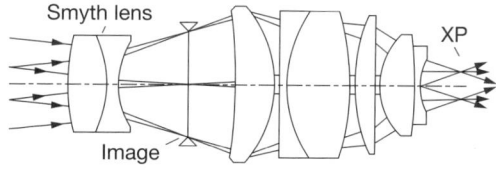

This eyepiece has two drawbacks: the lenses are especially large in diameter, and excessive spherical aberration of the principal rays in early models introduced a special kind of vignetting called the "**kidney bean effect**." This effect is described on page 78.

Selection of Interchangeable Eyepieces

Many spotting and most astronomical scopes are designed to accept **interchangeable eyepieces** as a means for changing magnification. A new scope may have one or more eyepiece(s) included. A potential problem for the beginning user of such instruments is determining which eyepieces from the large number of other commercially available types would be best to use with the scope. Here, some guidelines for eyepiece choices are suggested.

Modern scopes with interchangeable eyepieces usually have cylindrical sockets sized to accept eyepieces with 1.25-in (31.75-mm) or 2.00-in (50.80-mm) outside diameter cylindrical interface tubes. Hence, the first criterion for selection is to ensure that the eyepiece will fit into that interface.

Many beginning astronomers start with a classic Newtonian scope, such as a 152 mm, $f/8$. Three eyepieces should be adequate as a starter set. First, might be one with a ~4.5(f/number) = 36-mm EFL ($M = 34\times$) and an AFOV of 50° to 60° for finding targets. The XP would then be ~4.5 mm and suited for observing low-contrast targets. This would match the younger eye pupil in low, but not very low light levels or that of an older user under brighter conditions. A Plössl or Erfle design would be an appropriate choice.

Next, one should have an eyepiece giving an XP of ~2.5 mm diameter for observing targets under optimum seeing conditions. In the Newtonian, the M would be ~60× and the EFL_{EP} would be ~20 mm. A 45° orthoscopic or a 50° Plössl eyepiece should work well here.

Finally, the set should have an eyepiece that gives an XP of ~0.8 mm for maximum resolution when splitting close double stars or observing planets. M would then be ~190× and the EFL_{EP} would be ~6 mm. A Plössl design should work, but a Nagler would increase the XPD.

Of course, additional eyepieces would enhance the utility of the scope and allow experimentation under various seeing conditions to find the best combination for maximum information gathering and pleasure.

Selection of Interchangeable Eyepieces (cont.)

A larger set of eyepieces used by one experienced amateur astronomer in a variety of scopes is shown here. They are all are shown at the same scale. Optical parameters and dimensions are listed in the following tables.

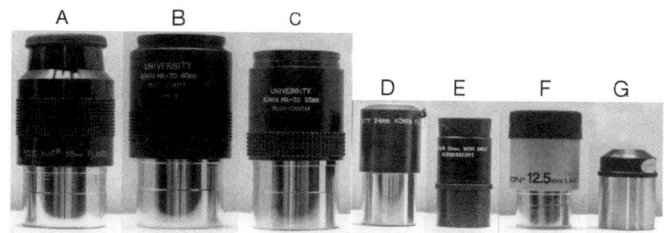

Eyepiece designation, EFL, and source	Type	XPD (mm)	D_{FS} (mm)	AFOV
A 55 mm TeleVue	Plössl	38	46	50°
B 40 mm Universal	König	17	47	70°
C 25 mm Universal	König	17	29	68°
D 24 mm Universal	König II	16	28	60°
E 20 mm Vernonscope	Brandon	—	22	65°
F 12.5 mm Orion	—	20	12	50°
G 6 mm Edmund	Orthoscopic	6	4.9	43°

Note: D_{FS} is the diameter of the field stop (see page 74).

Eyepiece designation, EFL, and source	Tube diam. (mm)	Length (mm)	Max. diam. (mm)
A 55 mm TeleVue	50.8	116	59
B 40 mm Universal	50.8	111	60
C 25 mm Universal	50.8	103	57
D 24 mm Universal	31.8	69	35
E 20 mm Vernonscope	31.8	59	32
F 12.5 mm Orion	31.8	67	39
G 6 mm Edmund	31.8	46	33

The Field Stop

Most binoculars and scopes have a ring-shaped diaphragm at an internal image to sharply define the edge of the field. This is called the **field stop**. In binoculars and scopes with permanent eyepieces, the field stops are typically attached to the housings, prism mounts, or into fixed parts of focusing eyepieces. If a reticle and/or a field lens is provided at the image, the field stop might be built into its cell. In spotting scopes with lens erectors, it can be at either image. The field stop is typically at the second image in US riflescopes.

The diameter D_{FS} of the field stop in millimeters is related to the objective and eyepiece EFLs (f_{OBJ} and f_{EP}) in millimeters and the real or apparent fields of view in degrees as:

$$D_{FS} = 2(\tan \alpha)(f_{OBJ}) = 2(\tan \beta)(f_{EP})(M_E), \text{if at } 1^{st} \text{ image,}$$
$$D_{FS} = 2(\tan \alpha)(f_{OBJ})(M_E) = 2(\tan \beta)(f_{EP}), \text{if at } 2^{nd} \text{ image,}$$

where M_E is the erector magnification (unity for prisms).

Scopes with interchangeable eyepieces usually accept eyepieces with 1.25-in (31.75 mm) or 2.00-in (50.80 mm) diameter interface tubes. The field stops are built into those eyepieces and are no larger than ~28 or ~46 mm. The maximum RFOV and AFOV, in degrees, are then

$$2\alpha_{1.25} = 1604.3/[(f_{OBJ})(M_E)] \quad \text{and} \quad 2\beta_{1.25} = 1604.3/f_{EP}$$
$$2\alpha_{2.00} = 2635.6/[(f_{OBJ})(M_E)] \quad \text{and} \quad 2\beta_{2.00} = 2635.6/f_{EP}$$

Here is the way to calculate the RFOV and AFOV of such a scope. Given: $f_{OBJ} = 100$ mm, $M_E = 2.0$, and $f_{EP} = 55$ mm so $M = (100)(2.0)/55 = 3.64\times$. The 50.8-mm barrel eyepiece is designed for 50°, so $D_{FS} = (2)(\tan 25°)(55.0)$ should be 51.29 mm. This exceeds the maximum, so D_{FS} is set at 46 mm. Then, $\beta = \arctan\{46/[(2)(55)]\} = 22.69°$ and $2\beta = 45.4°$. Finally, $\tan \alpha = \tan \beta/M = 0.418/3.636 = 0.115$ and $\alpha = 6.56°$, so the total RFOV is 13.1°.

Parallax

Parallax may be seen in binoculars and scopes when images of nearby targets are superimposed on fixed references such as a reticle pattern. If there is no such reference, parallax is not observed. If the target is at an infinite distance and the reticle is in the plane of the objective's infinity focus, the image and the reference will seem to move together as the eye moves laterally within the XP. If the target is close to the optical system, its image will be axially displaced toward the eyepiece from the plane of the fixed reference, and relative lateral motion will be observed between the target and reference if the eye decenters. This is parallax.

This effect can cause pointing errors when the instrument is used as an aiming device on a rifle or another weapon. The magnitude of the error depends on the size of the XP, which, in turn, depends upon the D_{EP} and M. Larger pupils lead to larger parallax errors.

These errors can be reduced in a fixed-focus instrument during manufacture by focusing the objective at, for instance, 100 m. As long as the target is at or near that distance, parallax will be essentially nil. Aiming at a target at 500 m without readjusting the objective focus could, however, result in a sighting error.

An obvious solution to this parallax problem would be to optimize focus of the objective at the reticle for each target distance. Spotting scopes and binoculars with center or internal focus allow focus adjustments. In these cases, a way to determine when best focus is achieved is to move the eye in the XP to determine whether parallax can be observed. If so, the focus is not correct and refinement is needed.

Note that the usual diopter movement to correct focus of the eyepiece is still needed in order to correct refractive errors of the eyes and achieve best imagery of both the reticle and the target.

Light Transmission

In this section, we show how to estimate light transmission T through a binocular or scope if the specifications or examinations of the instrument reveal the pertinent details about its construction.

The following factors reduce light transmission:

- **Absorption losses** along the glass path at T_A/cm,

- **Fresnel reflection losses** at refracting surfaces at T_F/surface,

- **Reflection losses** at 1st and 2nd surface mirrors at T_R/surface, and

- **Obscuration** at beam area obscured/total beam area at point of obscuration.

For the purposes of this Field Guide, the following values for the various T factors are assumed:

- T_A for all optical glasses = $(0.998)^{\Sigma L}$, L in cm;

- T_F for N uncoated refracting surfaces = $(0.960)^N$;

- T_F for N MgF$_2$ coated surfaces = $(0.987)^N$;

- T_F for N multilayer dielectric (MLD) coated refracting surfaces = $(0.993)^N$;

- T_R for N 1st or 2nd surface-protected aluminized mirrors = 0.900^N or 0.88^N, respectively;

- T_R for N TIR 2nd surface mirrors = 1.000^N;

- T_O for obscuration determined for each occurrence;

- Total system transmission is the product of all applicable factors.

If the design or a scale diagram of the optical system is available, the glass path and number of refracting and reflecting surfaces can easily be determined. Otherwise, the number of air-to-glass interfaces in an objective or eyepiece can usually be determined by shining a light (such as an attenuated laser pointer) into the lens aperture and counting the reflections. It may be possible to count surfaces in lens erecting systems this way, but it is more difficult because of attenuation.

Light Transmission (cont.)

Cemented Porros have two MLD-coated surfaces, four TIR reflections, and a glass path of $4A$. Cemented Abbe-Königs have two MLD-coated surfaces, two TIR reflections, and (in the symmetrical form) a glass path of $5.2A$. Roof Schmidts have four MLD-coated surfaces, four TIR reflections, and one 2nd aluminized surface. Their glass paths are $4.62A$.

The glass path is estimated as follows: Assume that the objective is $f/4$ so f_O and f_E are known from D_{EP} and M. Objective thicknesses are ~0.25 times D_{EP}. Eyepiece path lengths are ~$0.5f_E$ (Kellner), ~$0.8f_E$ (RKE), ~$0.75f_E$ (Plössl), and ~$1.32f_E$ (Erfle). Apertures A of Porros are ~$0.90D_{EP}$, of Abbe–Königs are ~$0.70D_{EP}$, and of Roof Schmidts are ~$0.55D_{EP}$.

To illustrate transmission estimation, consider this 7×50 system with cemented BaK4 Porro prisms ($n = 1.569$). Dimensions are in millimeters and follow the above guidelines. Prisms are shown as **reduced tunnel diagrams**.

A reduced tunnel diagram shows prism axial lengths divided by n so that rays pass through without refraction.

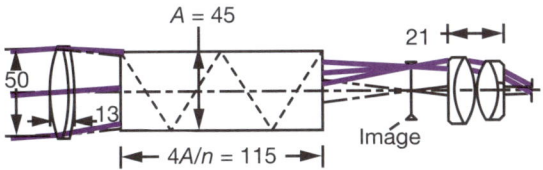

Actual glass path: 21.4 cm; MLD-coated surfaces: 8; and TIR surfaces: 4. Then, T is: $(0.998^{21.4})(0.993^8)(1.000^4)$ or $(0.958)(0.945)(1.000) = 91\%$.

Similarly, T for a WWII 7×50 binocular with MgF$_2$-coated surfaces, air-spaced BK7 Porro prisms, and glass path assumed to be the same as above is estimated as 73%. This agrees precisely with 1947 US Naval Ordinance measurements of that instrument.

Vignetting

Classical **vignetting**, illustrated schematically below left for a simple Keplerian scope comprising only an objective and an eyepiece, occurs when a portion of a light beam falls outside the usable aperture of some downstream lens and is lost.

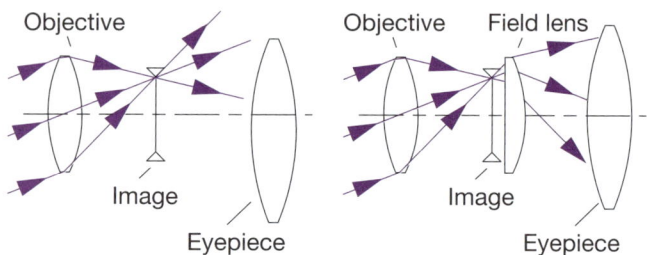

Vignetting to ~50% of a beam from the edge of the field is often tolerated in visual systems because the eye may not notice the image luminance decreasing there. The optics then can be smaller than if the full beams were transmitted. Usually, the full beam should be transmitted over the central 70% of the field.

The figure at right above illustrates how adding a positive field lens near the internal image can redirect the beam into subsequent optics—thereby reducing vignetting.

The "kidney bean effect" mentioned on page 71 occurs because the principal rays for different field angles in an eyepiece intercept the axis at different distances along the axis; i.e., they display spherical aberration of the XPD. Depending on the position of the eye with respect to the eyepiece, part of the field will not be visible; a dark shadow, shaped like a kidney bean, will be seen. It will move as the eye moves to scan the field. Early Nagler designs had this problem, but later designs reduce or eliminate it.

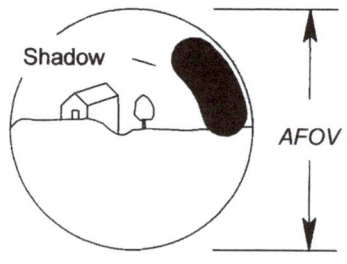

Stray Light

The term **stray light** refers to light from a source inside or outside the FOV of an optical system that reaches the image sensor (here, the eye), but detracts from—rather than contributing usefully to—the image. Stray light may appear as ghost images superimposed on the main image. It may also take the form of diffuse illumination (called **veiling glare**) superimposed on the entire image or parts thereof that reduces image contrast and MTF.

The chief causes of stray light include:

- light reflected or scattered from the inside surfaces of housings, cells, or lens rims that ultimately finds its way to the eye,

- illumination of the sensor by rays bypassing the optics,

- light scattered or diffused from dirty or moisture-laden refracting or mirror surfaces,

- multiple reflections within lenses, and

- internal reflections within prisms (especially if the prisms are not made slightly oversize with respect to the minimum-required clear aperture).

Measures that reduce stray-light problems include:

- efficient antireflection coatings on refracting surfaces,

- internal mechanical **baffles**, **glare stops**, and field stops,

- texturing, threading, and blackening of internal mechanical surfaces,

- maintaining cleanliness of the optics, and

- external lens hoods.

A glare stop is a ring-shaped diaphragm located strategically in a binocular or scope to reduce stray light, usually at an image of the AS (e.g., inside the erecting lens system of a rifle scope).

Light Baffles

The objective of a refracting scope or binocular admits light into the instrument from many directions outside the conical RFOV. For example, as shown below for a riflescope, sunlight can enter from above and illuminate the bottom interior surface of the housing. If this surface is rough and black, much of this light is absorbed, but some is scattered into the cavity. A portion may pass through the optics and enter the eye.

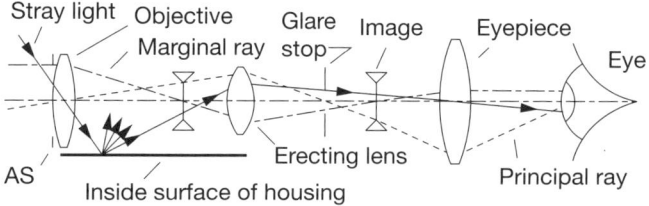

If the housing is appropriately shaped, a portion of its interior surface might have machined or cast threads (see top of second figure). These threads act as absorbing barriers against refracted stray rays such as #1′.

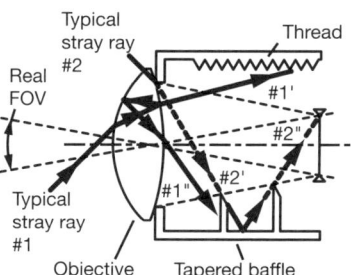

Stray ray #1″ is formed by dual internal reflections inside the objective. Its ill effects are prevented by a baffle such as the one shown at the bottom of the second figure. That baffle is usually inserted as a separate part into the housing. Note that the sharp corners of the two ring-shaped diaphragms extend inward to the dashed line connecting the objective aperture with the field stop at the image. The central conical volume thus formed is kept clear for passage of the image-forming beam.

The first vane is positioned at the intersection of stray ray #2 and the dashed marking the clear zone. The second vane is at the intersection of ray #2″ and the same dashed line. More complex baffling arrangements, including exterior lens hoods, are needed in other instruments.

Light Baffles (cont.)

Reflections within some prisms can introduce stray light into the imaging system. The Porro prism is an example. Rays reflected close to the first 45° corner will reflect internally from the hypotenuse and create stray light in the system. Grooves cut across each hypotenuse will block them without affecting desired rays.

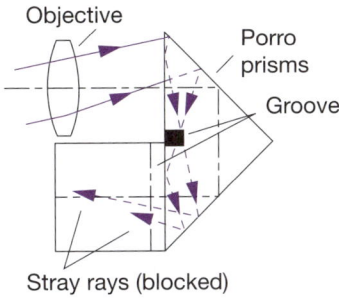

The image below illustrates a Schmidt–Cassegrain astronomical scope with two tubular baffles that are designed to block all stray light entering the system from outside the field of view (the colored arrow is a ray at the edge of the FOV). The baffles also block rays reflected or scattered from the interior surface of the housing. Note that one or both of the baffles may increase the central obscuration. The primary baffle may be cylindrical, but the secondary baffle is generally conical. Similar baffles are used in Gregorian-type systems.

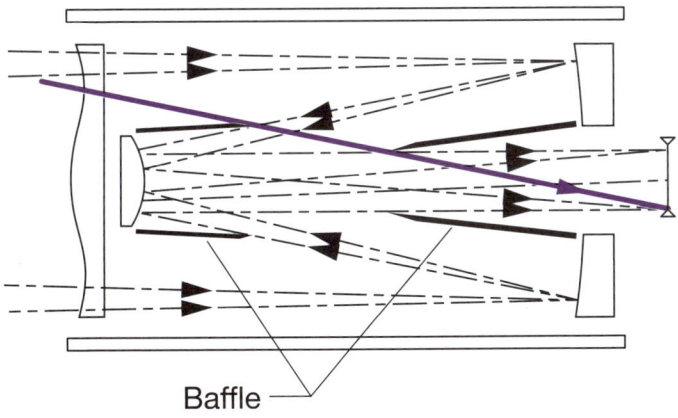

Reticles

A **reticle** is provided in scopes used with weapons to define
the point in the FOV where the bullet trajectory intersects the
aiming point if the optics and weapon are properly aligned to
each other. The reticle may comprise very fine wires in the form
of a cross or a glass plate with the pattern engraved or deposited
thereon. Shown here are some common patterns:

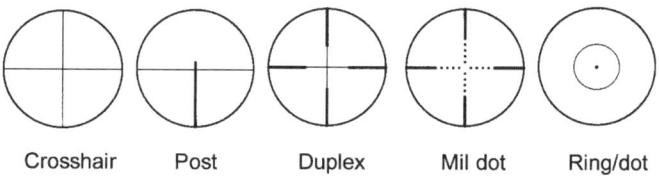

Crosshair   Post   Duplex   Mil dot   Ring/dot

The best pattern for a particular application depends largely
on the target visibility and scene luminance. If the field is
cluttered by foliage or is barely visible in dim light, heavy lines
are needed in order to be seen. Military assault riflescopes
often use a simple wide post for maximum aiming/firing speed
and visibility in low light. Some reticles can be illuminated. In
competitive long-range target shooting, fine lines are needed for
accuracy.

For both the target and reticle pattern to appear in focus, the
pattern must lie in an image plane. In a riflescope, the reticle
may be located at either image plane (see page 17). In the US,
they usually are at the eyepiece focus.

Reticles cannot be used in a Galilean scope or field glass because
there is no internal real image plane.

Zeroing means bringing the optical
LOS into coincidence with the point
of impact of the weapon. This is done
by observing the strike patterns of
multiple groups of test shots and
then adjusting elevation and windage
of the optics to coincide with the
centroid of the pattern.

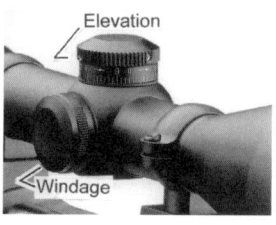

Variable-Magnification (Zoom) Systems

Zoom optics are provided in the objectives of a few binoculars and many spotting and rifle scopes to allow continuous variation of magnification. The FOV in object space varies inversely with magnification. Typically, the user might want a large field with low magnification to acquire a target and a narrower field with high magnification to examine target detail.

In zoom designs, at least two lenses are moved along the axis to change M, as indicated schematically here for an objective.

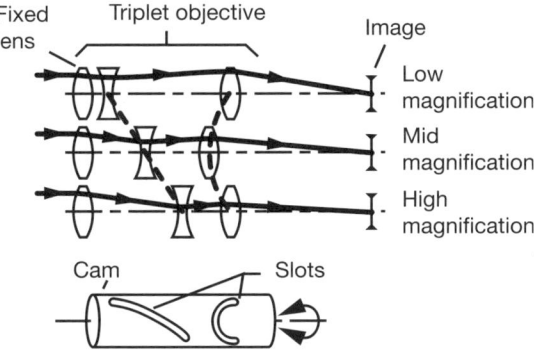

The thick dashed lines show the lens motions that change magnification while keeping the image in focus. These motions are typically driven manually by rotation of a zoom ring on the instrument housing that turns a cam to drive the lenses along the axis.

A zoom erecting system is also used in some designs. It appears similar to that shown above, but the object for the fixed lens is at the objective focus.

It is important for the lens motions not to disturb the line of sight, especially when the direction of cam rotation reverses. This requires tight tolerances on motions of the moving mechanical components.

To maintain good optical performance without undue complexity (and related higher cost), the magnification change should be limited to a factor of 2 or 3.

Variable-Magnification (Zoom) Systems (cont.)

The zoom function is designed into the eyepiece of some spotting scopes and binoculars. This is a convenient way to change magnification without requiring refocusing; however, a simple zoom lens can cover only a limited range of magnifications. One such design is shown schematically below. It uses a negative Smyth field lens (see page 71) as one moveable component.

To increase the range of magnifications, more complex systems would be required.

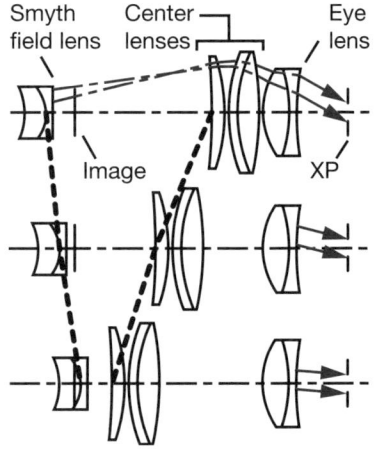

Some manufacturers offer a set of interchangeable zoom eyepieces so that a large range can be covered continuously with just a few eyepieces.

A variation on the theme of zoom instruments is the dual-position zoom that provides good imagery only for two extreme focal lengths and has a mechanism to switch from one to the other. For example, this feature is offered in two binocular configurations of the Leica "Duovid": a $8 + 12 \times 42$ and a $10 + 15 \times 50$. To toggle from one magnification to the other, the user rotates a ring on each eyepiece by one-quarter turn. To revert the settings, the rings are returned to their original positions. This mechanism for magnification change is not very convenient.

Other dual-position zoom designs use a single lever on the back of the housing to accomplish the magnification change. This might be more convenient than the rotating ring approach since the lever can be moved by one finger without removing the binocular from the eyes. An example is shown on page 21. The lever is directly in front of the focus ring.

Image Stabilization Techniques

Muscular tremors and external vibrations may make it difficult for a user to hold a scope or binocular steady enough to achieve maximum visual performance. The angular excursions of a handheld scope or binocular are typically sinusoidal at about 9 Hz and have amplitudes of about ±0.25°. **Stabilization** allows a handheld high-powered instrument to give image quality comparable to that of the equivalent nonstabilized tripod-mounted instrument. Because the object of interest is typically at a large distance, only angular motions are of concern. Instrument translations are not compensated.

We consider these three types of functionally equivalent internal mechanisms have been used. The first type has a compensator lens (or lenses acting as a group) that is moved laterally in orthogonal directions by actuators as required to deviate the moving image and make it appear fixed (see sketch below). Drive signals are provided by a microprocessor in response to motions sensed by orthogonal accelerometers. Battery power is needed.

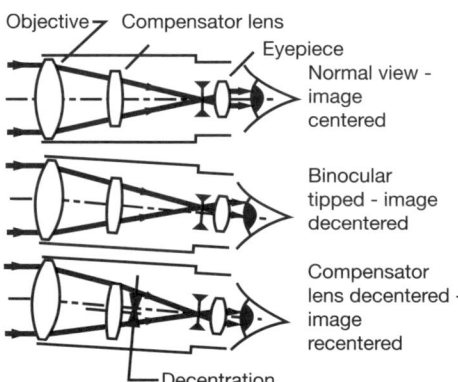

In a second approach, erecting prisms are suspended in a two-axis **gimbal mount** or on **flexures** to inertially stabilize the LOS in response to small instrument motions. No electrical power is needed. A design variation incorporates motor-driven **gyroscopes** into the prism suspensions.

Image Stabilization Techniques (cont.)

The third type incorporates a variable **optical wedge** into the optical system. This component has two thin windows connected by a bellows to constrain a fluid. In response to sensor signals, one window is tilted in orthogonal directions by actuators to deviate the beam and restore image centration.

Another way to stabilize the LOS of a scope or binocular involves an **external stabilizer** that steadies the entire instrument rather than just the image. One such stabilizer that has been used for over 50 years is sketched below. It has a "pod-shaped" housing measuring ~71 mm in diameter and ~114 mm in length and weighing ~1 kg. It attaches to the bottom of the visual instrument through an adapter and is powered by a separate battery. The stabilizer axis is parallel to the LOS. Inside the unit, two orthogonal gyroscopes spin at high speed and stabilize pitch and yaw motions of the assembly.

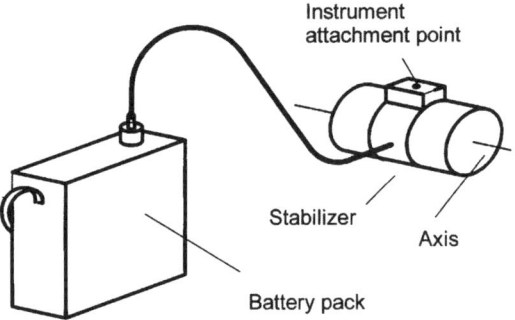

The dynamic angular ranges of all of these image-stabilizing devices are limited, as are their responses to high-frequency vibrations.

Rangefinding Techniques

Some special binoculars and riflescopes provide means for estimating target ranges by active or passive (**stadiametric**) rangefinding techniques. The active type projects a pulsed beam of infrared **laser** or LED light to the target. The optics are indicated conceptually here.

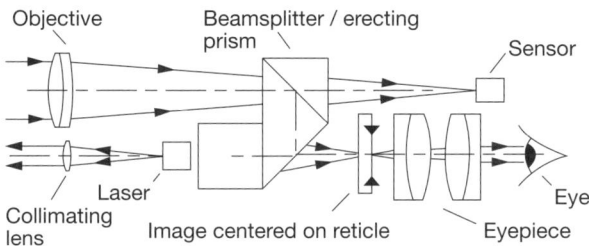

A portion of the beam reflected by the target is collected by the objective lens, transmitted by the beamsplitter, and sensed by a detector. Ambient light reflected by the target reflects from the beamsplitter to the eyepiece and produces a visual image. The range is determined electronically from the time interval required for a light pulse to travel to and back from the target. It is displayed digitally within the eyepiece FOV.

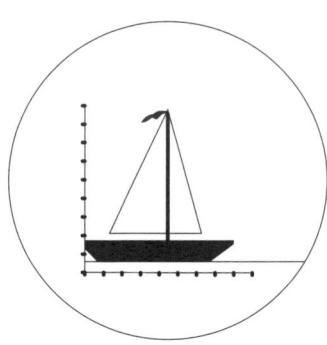

The stadiametric approach is simpler but less accurate. Two orthogonal lines divided into equal angular parts are on a reticle. Target length or height must be known or approximated. In the former case, the instrument is pointed so that the image's left end touches the vertical line. The subtended angle of the target's length is read from the horizontal scale, and the range is calculated as length/angle. If the angle is in radians, the range is in the same units as the target length.

Overall Size of a Binocular

The physical diameters and f/numbers of the objectives, the prism geometry, the objective cell and housing wall thicknesses, the apparent field, the XPD, and the IPD are major factors in determining length, width, and thickness of a binocular. The following schematic shows a top view of one-half of a binocular with Porro prisms. Many of the factors listed above that affect length and width are designated.

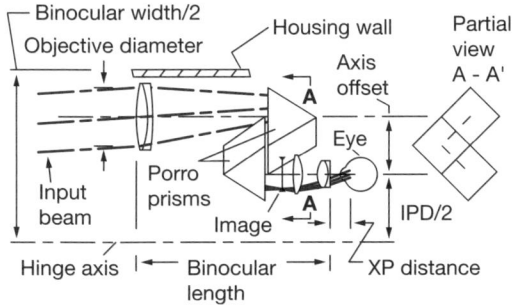

Beams from distant objects converge as they approach the prisms, so the prism face width A is smaller than the objective aperture. All prism dimensions are proportional to A. For example, the axis offset in a Porro system is $1.41A$. For a 30-mm EFL objective, $A = \sim 27$ mm, and the offset is ~ 38.0 mm. IPD is usually ~ 52 to ~ 75 mm.

Binocular length is determined by the objective and eyepiece EFLs and physical lengths, the path-lengthening effect of the prisms [equal to $(n-1)t_A/n$ where t_A is the path length in the glass ($t_A = 4A$ for Porros) and n is the refractive index], and any axial path reversals in the prisms.

The physical length of a symmetrical Abbe–König prism is $3.46A$, and its glass path length is $5.20A$, whereas those dimensions are $1.21A$ and $4.62A$ for a roof Pechan prism. A significant portion of the path within the latter prism is directed laterally. Binoculars using roof Pechan prisms can be somewhat shorter than those with Abbe–König prisms.

Overall Size of a Binocular (cont.)

Major factors contributing to binocular thickness are objective aperture and lateral dimensions of the prisms, as indicated in the figure below. The thicknesses of objective cells and the housing walls also contribute to this dimension.

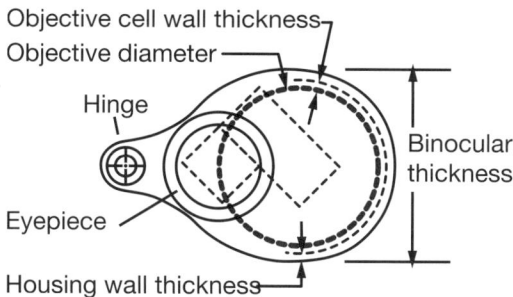

The thickness of an in-line binocular is usually determined by the objective diameter and its cell and housing wall thicknesses.

In applications in which depth perception is not essential, the width of a binocular can be significantly reduced by locating the objective axes that are closer together than the eyepiece axes, as shown in the following sketch and the photograph on page 20.

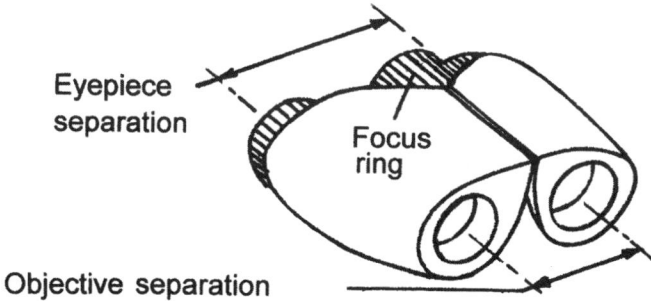

Porro prisms are usually used to minimize costs. In this application, they are referred to as **"inverted Porro" (IP)** assemblies. The instrument's width is slightly larger than the IPD plus the eyepiece outer diameter.

Weight of a Binocular

Significant factors that contribute to the weight of a binocular include objective and eyepiece dimensions, the number of lenses, prism configurations and sizes, housing configurations, and material choices. Special features, such as LOS stabilization, variable magnification, an internal compass, a laser rangefinder, a reticle, laser filters, or exceptional structural ruggedness (as might be required for military applications) add weight.

Many of these factors are interrelated. For instance, the weight of the optics in an eyepiece is determined by its diameter, the number of lenses, the element configurations, and the properties of the glasses. These factors depend on image quality requirements. Eyepiece aperture depends strongly on the XP diameter, the XP distance, and allowable vignetting.

The weights of the prisms depend on their configurations, face widths A and glass densities ρ. Porro, symmetrical Abbe–König, and roof Schmidt types weigh about $2.57\rho A^3$, $3.72\rho A^3$, and $1.80\rho A^3$, respectively. Because the optical path lengths inside the prism assemblies are different, the various prisms will be located at different distances from the objectives in the convergent beams and their apertures will then be different.

To indicate the magnitude of the weight variation for different size binoculars, the weights of 91 Porro prism models and 118 roof prism models with objective apertures ranging from 15 to 80 mm as specified by nine well known manufacturers were tabulated. The selected models were conventional in configuration, i.e., did not include ones with special features, such as stabilization, range finding, or mounts.

Weights for units with the same apertures in each group were averaged. These averages were plotted to create the graph shown on the next page. The solid line is an approximate spline fit to the 28 points plotted.

Major weight differences between Porro and roof prism designs cannot be observed from the averages, although some differences do exist for individual models in each size category.

Weight of a Binocular (cont.)

Note that, in this sampling, models with roof prisms are not available with objectives larger than 56 mm.

Weight (g)

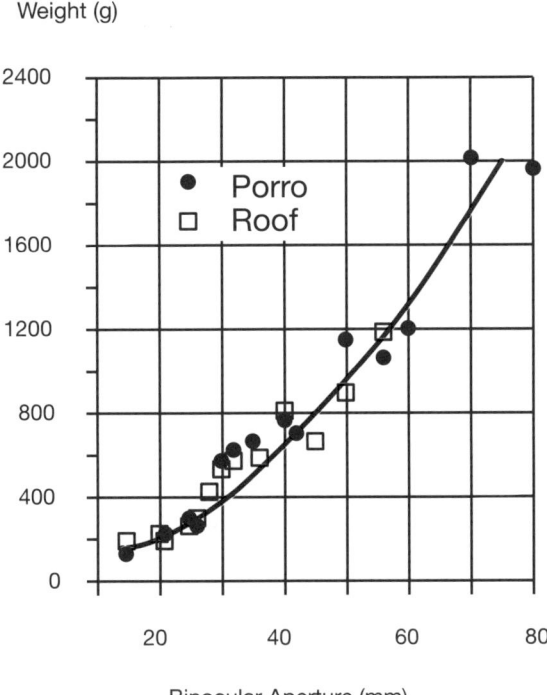

Binocular Aperture (mm)

Studies have indicated that excessive weight of a handheld unstabilized binocular or scope causes user fatigue and hinders the user's ability to track rapidly moving objects such as military targets. They also indicate an upper limit of about 2000 g to be appropriate for extended use. Other studies show that increased instrument weight, within this rather arbitrary limit, does not materially increase hand shake or muscular tremble.

To first-order approximation, one would expect that spotting scope weights would be about one-half those of optically equivalent binoculars.

Ergonomics

The synergism between user and equipment is referred to as **ergonomics**. In the present context, it pertains to how easy it is to hold the binocular or scope and to operate all controls, such as IPD, focus, and diopter settings. How the instrument performs, how it interfaces with the eyes, and the degree to which it meets the peculiar needs of the application also are concerns.

  This figure compares 1968 and 2000 designs for a Zeiss 8 × 56 binocular with Abbe–König roof prisms. The length was reduced by ~17%, primarily by replacing the doublet objective with a four element, air-spaced telephoto objective design similar to that shown on page 35.

The shapes of the housings in the Zeiss design evolved from connected cones and cylinders to a smoothly curved contour that is easier to hold. The **focus ring** was moved forward for easier access by the fingers when they are wrapped around the binocular near its center of gravity for best balance and stability. The **diopter adjustment** was associated with the focus ring rather than being on one eyepiece. Internal focusing resulted in a sturdier and sealed structural design. The use of lead-free, high-index glasses and optomechanical design optimization helped reduce weight by ~20%, even though the apparent field was increased from 50° to 60°.

Ergonomic design improvements provided in other modern binoculars include eyepieces with increased XP distances, **eyecups** that retract or can be removed to accommodate spectacles, and eyepiece contours that minimize interference with the nose or brow. The latter improvement may conflict with a requirement for a wider FOV or longer XP distance, either of which calls for larger eyepiece apertures and, hence, larger eyepiece external diameters.

The feel of the surfaces of a binocular or scope's housings contributes to comfortable use. Many models have slightly resilient rubber or plastic covers that improve the grip as well as provide protection from bumps, abrasion, and weather.

Ergonomics (cont.)

Strategically placed ridges and depressions may provide convenient grips for fingers and thumbs, but these features may not fit both large and small hands.

In an ergonomically successful instrument design, all mechanisms will function easily yet retain adjustments under adverse environments. For example, binocular hinges should move smoothly without requiring excessive force at extremely low temperatures, while providing enough stiffness to hold the IPD setting when the binocular is held in one hand and moved rapidly from one orientation to another at higher temperatures.

Optical performance of binoculars or a scope is of prime importance to the user. Images should be sharp and have high contrast to the edge of the FOV. Color fringing at the edges of an image may be particularly annoying. This usually results from slightly varying magnification for different colors of light in the eyepiece design and increases with field angle. A finely tuned lens design using low-dispersion optical glasses will minimize this aberration. The magnification of a binocular or scope should be nearly constant over the FOV. If not, either of two types of distortion may exist, as discussed on page 54.

To minimize eye strain while using a binocular, the magnifications of the two scopes should be equal within 2%. The rotational orientation of the two images about each axis should match within ~1°. These specifications are determined during manufacturing. Assembled instruments do not have user adjustments for magnification or image rotation. The focus of each eyepiece should be adjusted to match the user's diopter setting within ~0.25 D.

Environmental Considerations

Important environmental considerations for binoculars and scopes are **temperature**, **sealing**, and **mechanical shock/ vibration**. Manufacturers of nonmilitary equipment usually apply their own proprietary specifications. Testing is at their discretion and is usually done only on expensive models. Military instruments are designed, made, and tested to military specification levels. In the US, MIL-STD-810, *Environmental Test Methods and Engineering Guidelines* applies. Elsewhere, ISO 9022, *Methods for Testing Optical Components and Optical Instruments* applies. These documents sometimes serve as guidelines for nonmilitary applications.

The temperature of an optical instrument is rarely uniform. Misalignment, defocus, and decrease in quality of the image may result from differential thermal expansion. Typically, during use at −51 °C to +49 °C, performance should be as specified. Damage should not result from exposure to temperatures of −61 °C to +71 °C.

Protection against intrusion of water into an instrument is often specified using **International Ingress Protection (IP)** codes, normally intended for electrical enclosures. Tests for exposure to splashing water and extended immersion to depths of 1 m or more are defined in these codes. Detectable internal water after testing is cause for rejection because this can fog optics and may cause failure of structural adhesive joints or coatings.

Mechanical shock exposures typically require design for and testing to performance standards per the end item specification. This requires evaluation after a minimum of three consecutive shocks equivalent to a drop of 1.2 m applied along each of the three axes.

Vibration testing typically requires exposure to a particular range of accelerations at specified frequencies for given durations without loss of performance resulting from component misalignment.

Housing Design

The traditional configuration for scopes and each half of a binocular has a rigid metal **housing** that serves as the main structural member. Interfaces are machined into the housing for the optics (lenses, prisms, etc.). Optics are usually held in cells or subassemblies. This construction principle is illustrated by the riflescope shown on page 33. The tubular housing extends from the objective to the reticle. The objective is inside a cell that is attached to the forward end of the housing. The focusing eyepiece subassembly is attached to the rear end of the housing. Joints may be sealed, usually with **O-rings**.

In typical contemporary binoculars and scopes, the housings are cast or molded, machined locally inside and out, and (usually) covered with a protective layer of rubber or plastic. Leather coverings are no longer used because they are susceptible to damage from fungus in warm, humid environments.

In the military binocular shown on page 26, the housing material is fiber-reinforced polycarbonate, molded to provide internal and external features that would have to be machined if the housing were metal. Other instrument parts are frequently attached to the plastic housing by adhesive bonding. This type of construction is utilized in many optical instruments intended for consumer use to reduce manufacturing costs, but it renders maintenance or damage repair technically difficult and expensive. Structural damage may result from impacts that commonly occur in military and similar applications.

Except for the example just mentioned, most binoculars and scopes used for military, marine, or law enforcement applications have metal housings to ensure durability. Anodized aluminum is most common, but titanium and magnesium (with protective finishes) have also been used. Wall thicknesses can be smaller with metals than with plastics because they are stiffer than plastics. This helps minimize the weight of the instrument.

Binocular Hinge Mechanisms

The **hinge** in a binocular allows the IPD to be adjusted to the user's setting. It should function smoothly and retain the setting until intentionally changed.

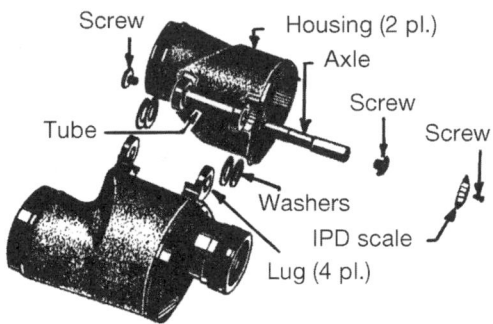

A simple and reliable classic design is shown in this exploded view.

A tube with a tapered bore is pressed into holes in the hinge lugs on the right housing. The lugs on the left housing slip over the right set of lugs with thin plastic washers between the adjacent surfaces. A lubricated tapered axle seats into the tapered bore. Screws are threaded into each end of the axle and tightened to squeeze both sets of lugs together to provide friction that holds the IPD setting. At assembly, the IPD scale is attached to the end of the axle, aligned to an index line engraved on one lug, and secured in place. In lower-cost designs, the tapered interface can be replaced by two axially loaded sleeve bearings that create friction.

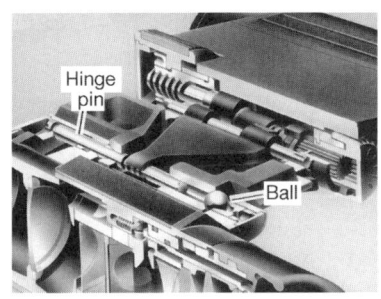

This hinge is one of two in the 8 × 20 binocular shown on page 63. A metal ball is clamped between conical seats to provide friction.

In the US Army's M19 7 × 50 binocular, a series of 11 lubricated O-rings are seated in grooves on a solid hinge pin locked to one scope. This subassembly is squeezed into a slightly undersized bore of a hinge tube locked to the other scope. The compressed ring-to-tube interfaces provide friction.

Binocular Collimation Mechanisms

A binocular must allow the user to look at a distant object for extended times without suffering eyestrain. This requires that the optical axes of the scopes be nearly parallel. Discrepancies are called collimation errors and are vertical (**dipvergence**) or horizontal (**convergence** or **divergence**), as shown here.

These errors are caused by misalignments of the objectives and/or the prisms. Means are provided in the design to bring such errors within tolerances during assembly and should allow realignment if the binocular is damaged.

The most commonly used mechanisms are **eccentric mounts** for both objectives (see second schematic). The lens cell and ring are turned with two wrenches while observing a distant target or the beam from a **collimator**. The beams exiting the eyepieces may be observed with two parallel scopes, one with a crosshair and the other with a reticle showing allowable image decentration. A test instrument of this type is shown on page 113.

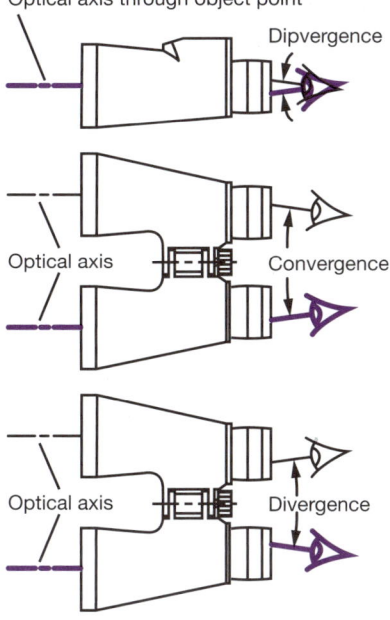

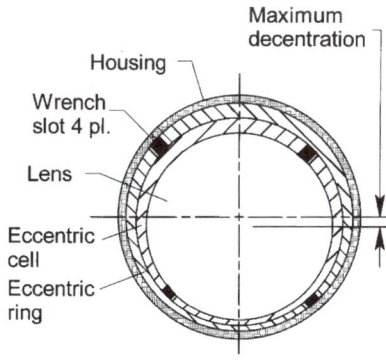

Binocular Collimation Mechanisms (cont.)

Typical tolerances on collimation errors that have been applied to military binoculars by the US Department of Defense are listed here. Note: convergence is not allowed.

	Minimum	**Maximum**
Dipvergence	–	15 arcmin
Divergence	5 arcmin	30 arcmin
Convergence	–	–

These tolerances apply at all magnifications and over the full range of IPD settings. Binoculars intended for consumer use frequently exceed these limits and may be measured at only one IPD. Higher-cost nonmilitary units may be held to tighter performance standards.

Users of binoculars that are suspected of having collimation errors are cautioned not to attempt repair without adequate test facilities. A test described in the literature involving observation of a selected distant point target with the eyes a meter or more behind the eyepieces and comparing where the images appear in the eyepiece apertures is *not* quantitative and is subject to misinterpretation.

Older binoculars without eccentric objective mounts may have adjusting screws in the prism mounts. These typically use push and pull screws acting in the interfaces between the prism mounting plates and the support points inside the binocular housing. Access to these adjustments is difficult, as is the adjustment process itself. Only experienced repair personnel should attempt realignment.

Some newer binoculars have no adjustments with which to correct collimation errors. They depend upon the accuracies of the individual optical and structural components and of their optomechanical interfaces to establish and preserve proper alignment.

Object Focus Mechanisms

Center focus mechanisms are commonly used in consumer binoculars that do not have a reticle to keep in focus. Most military and astronomical binoculars have individually focusable eyepieces (see page 63). The object focus function is then combined with the diopter adjustments to compensate for eye focus errors in this configuration.

One internal focus mechanism was briefly described on page 63; another is shown here. Turning the focus ring turns a short threaded shaft inside a rotationally fixed threaded collar and causes the collar to move axially. This motion is linked by two sliding focus shafts to the two objective cells. Axial movements of the objectives cause the images to move to the focal planes of the eyepieces.

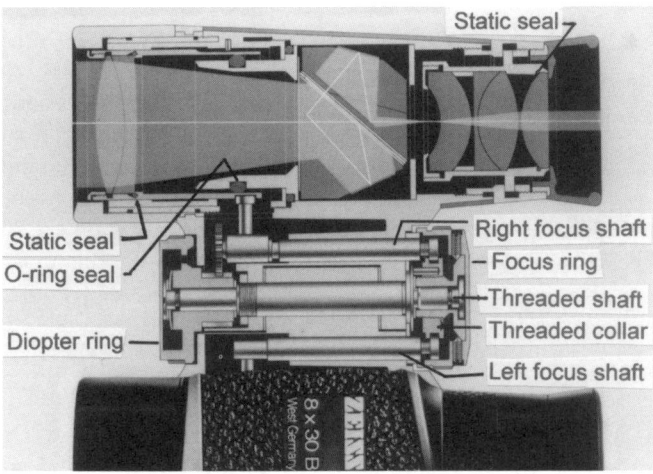

In use, after focusing the left scope, the diopter adjustment of the right scope (on top in the above figure) is accomplished by turning the diopter ring. This turns the gear on the associated focus shaft to bias the right objective location slightly with respect to that of the left objective, and brings both images into focus.

Note the O-rings that seal the translation motions of the objective cells: the outermost lenses are statically sealed.

Diopter Adjustment Mechanisms

A binocular with individually focusable eyepieces and no reticle gives the user the ability to adjust focus of the target image for each eye and simultaneously compensate for the eye's inherent near or far sightedness.

As shown here, rotating the focus ring moves the lens cell axially on the threads. If the lenses have centering errors from manufacture, rotation may affect collimation.

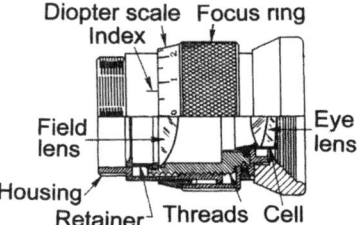

More sophisticated designs (as in the US Army's M19 binocular eyepiece shown here) slide the lenses without rotation. Rotation of the focus ring on its thread moves the lens cell axially. The pin slides into a slot to prevent cell rotation. The bellows seals the moving cell to the fixed housing, and the outer lenses are sealed statically to the cell at each end.

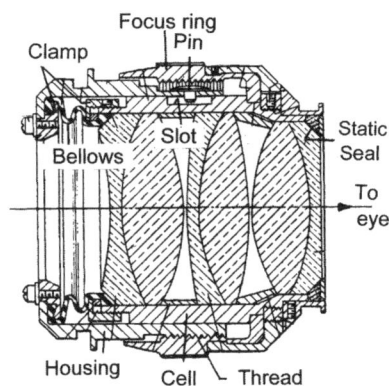

In a center focus binocular, the user usually focuses on the object of interest using one scope (which has no individual focus capability) and then uses a diopter adjustment mechanism in the eyepiece of the other scope to bias its focus to suit the associated eye.

The same capability is provided in some compact binoculars by mounting the objective lens of one scope in a focusable cell. The other scope is focused on the target, then the focus ring on the adjustable objective is turned to bring its scope into focus on the target. This allows each eye's defects to be compensated individually.

Sealing and Purging

Binoculars and scopes are, to varying degrees, able to survive exposure to moisture, such as high atmospheric humidity, rainfall, and temporary immersion in water.

Condensation forms inside a binocular or scope if the temperature drops below the dew point for the internal humidity. The interior must be dried out and the instrument sealed to prevent fogging. Purging with a dry gas is effective for removing water vapor.

Low-cost consumer items are generally least resistant to humidity, while those designed for military or law enforcement are most resistant. The eyepiece shown at the top of page 100 is not sealed, so fogging is a problem.

At the other extreme, in the eyepiece for a Swarovski riflescope (shown below), the outermost lens is statically sealed into its cell with an O-ring. Turning the focus ring translates the cell on a thread. The Quad-ring between the cell and the housing provides a dynamic seal for the focus mechanism. These features, combined with other O-ring static seals elsewhere in the scope, serve to isolate the scope's interior from the environment.

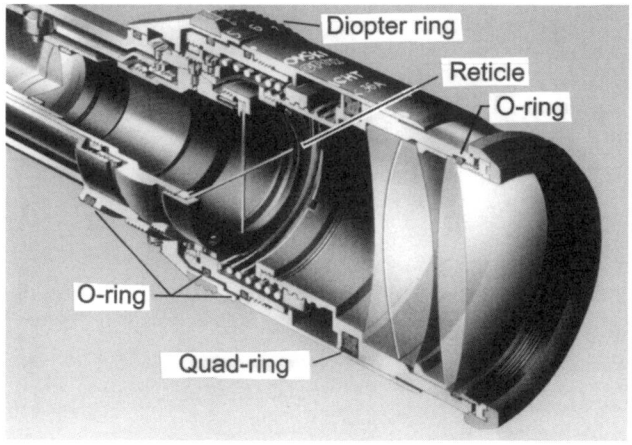

Sealing and Purging (cont.)

Note that a **Quad-ring** seals a moving part better in the long term than an O-ring does because it will not wear out as quickly from rolling friction during use. An even better solution is the rolling rubber diaphragm seal used in the US Army's M19 binocular, which has an extremely long lifetime.

Permeation of water vapor through seals is independent of the pressure differential between the inside and outside, so pressurizing an instrument provides little or no benefit. Water vapor entry is proportional to the difference in water vapor pressure. The worst-case vapor pressure difference is for a 100% relative humidity on the outside with a dry interior. The mass flow W of water through a seal is given by

$$W = VTR\frac{At}{L},$$

where VTR is the vapor transmission rate, A is the cross section area of the seal in m^2, L is the seal thickness in mm, and t is time in days. A typical value of VTR for the oft-used Buna-N O-ring seal is ~0.24 g-mm/m^2-day.

As a rough rule of thumb, to avoid fogging, the moisture content inside the instrument should be less than ~5 gm/m^3. In practice, it is difficult to completely dry out the interior of the instrument because materials such as adhesives and elastomers may emit water over time. Dry nitrogen is normally used for purging, since its water content is about 2 ppm. This corresponds to a dew point of -70 °C. Baking and purging is not a complete solution since the time and temperature for these operations are limited by both cost and potential for damage to the instrument's materials at higher temperatures.

Desiccants can be used as a means of absorbing residual water within the interior of the instrument after purging. They must be replaced over time, but this is not difficult if the instrument design provides easy access to the location of this material and for repurging.

Basic Photography Techniques

There are three main techniques for obtaining photographic records of the images seen through a binocular or scope:

- **Prime focus**. The camera's image plane is at the prime focus of a scope with its eyepiece removed.

- **Afocal projection**. The camera with its lens follows the eyepiece of the binocular or scope. The camera focus is infinity, and system focus is varied with the eyepiece.

- **Eyepiece projection**. An eyepiece projects a magnified image from the telescope on the camera focal plane with the camera lens removed.

The camera's auto focus, zooming, and image stabilization features can be used in these applications. A light-tight cover is needed between the binocular or scope and the camera in all cases. The camera also must be centered to the optical axis of the image forming system. When a digital camera is used, the afocal projection process is often called **digiscoping**.

The prime focus technique is most often used in astrophotography. The system's EFL and f/number are those of the scope's objective. A field-flattening lens may be needed in the camera if the scope's image surface is deeply curved.

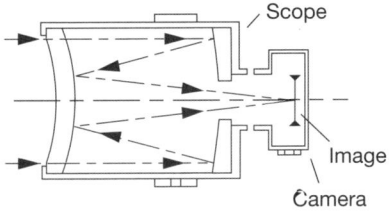

Key parameters of the scope/camera system when using the afocal projection technique are:

$$EFL_{AP} = M(EFL_C)$$
$$(f/number)_{AP} = EFL_{AF}/D_{EP}$$

where EFL_{AP} is the system's EFL, M is the magnification of the binocular or scope, EFL_C is the camera lens focal length, $(f/\text{number})_{AP}$ is the system's relative aperture, and D_{EP} is the system's EP diameter.

Basic Photography Techniques (cont.)

To record the entire image using afocal projection, the RFOV of the camera must be at least as large as the AFOV of the eyepiece. Further, the XP of the

eyepiece should be at the camera's EP.

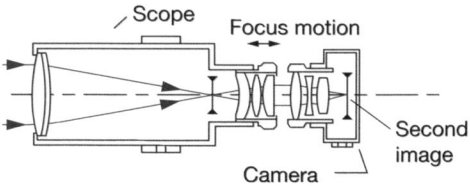

Usually, one does not know exactly where this EP is located, so experimentation with different camera lens-to-eyepiece separations (starting at the minimum) and examination of photographic images for uniformity of exposure, i.e., minimum vignetting, will indicate the best setting.

Eyepiece projection is used when maximum resolution is needed in lunar and planetary photography. The EFL and f/number of the scope/camera system are given by

$$EFL_P = [EFL_{SCOPE}][S/(L-S)]$$
$$(f/number)_P = EFL_P/D_{EP}$$

Here, EFL_P is the system EFL, EFL_{SCOPE} is the objective's scope EFL, L is the first image-to-second image distance, S is the eyepiece-to-second image distance, $(f/number)_P$ applies to the system, and D_{EP} is the scope's EP diameter. All dimensions are positive here.

The eyepiece used here should be of high quality. Microscope objectives are sometimes used (as illustrated). They also should be high quality.

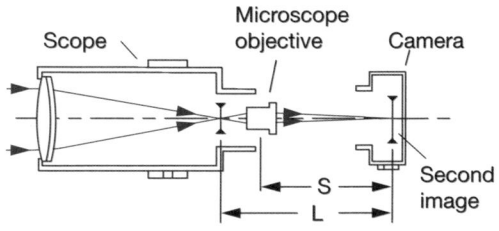

Interfacing the Camera

In prime focus photography, the camera body, without its lens, is supported so that the image plane of the scope coincides with and is parallel to the plane of the film or digital focal plane sensor. This usually involves the use of some kind of adapter that interfaces at its forward end with the scope and, at the aft end, with the feature on the camera to which the lens is normally attached. This feature may be a thread or a bayonet fitting customized for the camera in use. Because the sensor is typically located several millimeters beyond this feature, the camera body may need to protrude inside the back end of the scope. It is then best to use a camera-mounting adapter made by the scope manufacturer to avoid mechanical interferences with the scope.

Some astronomical scopes are designed to accommodate a **Barlow lens** in front of the image plane to increase the scope's focal length and, hence, the magnification. This auxiliary (negative EFL) lens also moves the image aft thereby making it more easily accessible to the camera.

In afocal projection photography, both the scope eyepiece and the camera lens act in series with a nominally collimated beam between them. A short adapter "tube" that slips over the eyepiece and the camera lens can usually be employed. With some cameras, the aft end of the tube can be screwed into a filter thread on the lens. This type of photography can also, in some cases, be used with binoculars.

For less critical applications, the camera can simply be held manually in contact with the eyepiece of the scope or binocular to photograph the image produced by that instrument. The image is centered by sliding the camera lens laterally on the rubber eyecup while observing the live image displayed in the viewfinder or LCD display.

In eyepiece projection photography, the camera is located some distance aft of the scope or binocular with an eyepiece or microscope objective located appropriately between them. This technique significantly increases the separation between the major optical components, so a different type of adapter is best used to connect them.

Interfacing the Camera (cont.)

As illustrated schematically below, a flat adapter plate or a lightweight beam can connect the scope to the camera for eyepiece projection, using screws passing through access holes or slots in the plate and engaging the tripod screw bushings normally found on the bottoms of practically all scopes, binoculars, and cameras. A support for the lens also is provided. The plate or beam is then attached to a tripod with another screw in the conventional manner. If possible, the tripod attachment point should be aligned with the center of gravity of the scope/binocular plus camera assembly for balance and stability reasons.

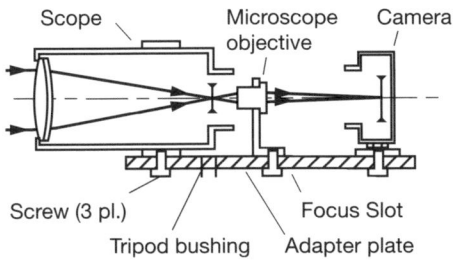

A different adapter configuration (not shown) uses brackets, rods, and clamps to attach the camera to the scope or binocular. By sliding the clamps along the rods, more adjustment flexibility can be provided than with the flat plate adapter. Some such adapters include hinges that allow the camera to be swung out of the way for access to the scope or binocular eyepiece(s) and conventional viewing of the target. Adding flexibility for use may also mean that the structure is more flexible from a vibration resistance viewpoint. These arrangements should be used with care to avoid image blurring.

Integral Cameras

Digital cameras can now be built into some binoculars and spotting scopes so that photographs may be taken of the scene observed through the eyepiece(s). Possible techniques include:

1. Adding a beam-splitting optical component (such as a modification to the erecting prisms) so that two images are formed simultaneously. One image is observed through the eyepiece, and the other falls on a CCD or similar focal plane sensor where the image is digitized. Single shot, multiple shot, or video outputs can be obtained. These signals are processed electronically for display, printing, or storage.

2. Adding a two-position mirror so that, in one position, the image is viewed through the eyepiece and, in the other position, the image is formed on the digital sensor. Simultaneous viewing and photography is not possible. The above outputs can be obtained.

3. A separate camera, with lens, LCD display, and digital output can be appended to the binocular or scope. The camera LOS is parallel to the visual LOS, so the images are essentially the same.

All of these techniques are more convenient than those described earlier in this section because the camera feature is always available for use, no adapters or other extra parts are needed, and alignment is accomplished during manufacture. The prime disadvantage is increased cost.

Describe here are two examples of integral camera instruments to illustrate this type digiscoping. The first is the Instant Replay binocular made by Bushnell (see upper figure on the next page). The second is the Zeiss Victory PhotoScope 85 T* FL spotting scope with integral digital camera (see lower figure, next page).

Note: a 65-mm aperture version of the latter instrument is also available as a somewhat smaller and lighter alternative to the described 85-mm version.

Integral Cameras (cont.)

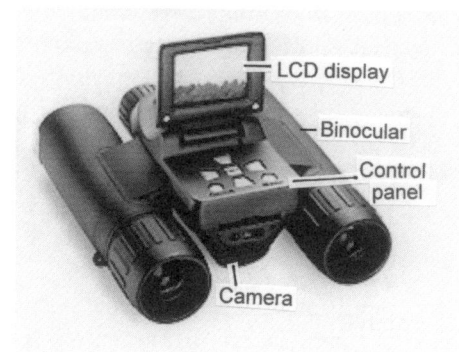

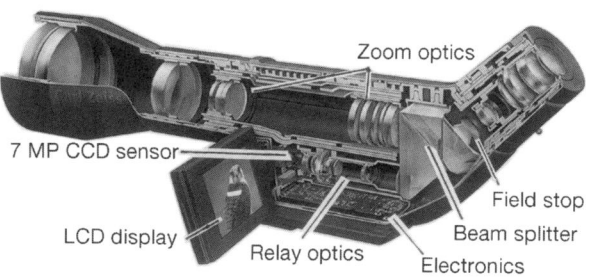

Key specifications for these instruments are as follows:

Company and model	*M*	Aperture (mm)	Erecting system	Weight (g)
Bushnell Instant Replay	8×	30	Roof	–
Zeiss Photoscope 85	15–45×	85	Lens	2900

The success of these and other digiscopes may lead to increased future usage in nature study, sports observation, and military/law enforcement applications.

Protection and Cleaning of the Instrument

High-quality binoculars and scopes are precision instruments. They need respect and care in order to perpetuate their performance. Exposure to extreme high or low temperatures, shock, vibration, blowing sand or dust, or excess moisture can disturb optical alignment, degrade the image, lead to eyestrain, or introduce contaminants that will degrade seeing through the optics.

Many contemporary instruments have resilient coverings on their housings and other exposed parts to attenuate bumps and minimize scratches during use. Carrying cases are usually provided for safe keeping of the equipment when it is not in use. Most of these instruments have protective covers for the exposed optics (objectives and eyepieces) as well as covers for openings left whenever interchangeable parts have been removed. If such devices are available, they should be used.

It is advisable to periodically clean exposed mechanical surfaces, such as housings, eyepieces, lens covers, mechanisms, etc. They should first be blown (with canned compressed air) and/or brushed free of dust (with a camel-hair brush). A soft cloth moistened sparingly with clean water can be applied to, hopefully, remove stains, fingerprints, or dirt spots from metal and plastic parts.

Optical surfaces require special care during cleaning. They, too, should be blown or brushed free of dust and then wiped *very gently* with lens tissue. The tissue should be folded into a small, elongated pad and held with fingers or tweezers at the opposite end or wrapped loosely over a cotton swab to avoid adding new contaminants to the surface. The tip of the pad should be moistened with a few drops of camera lens cleaner (not eyeglass cleaner). Wiping should follow a spiral path from center to edge, and the tissue replaced before wiping again. Lens cleaner should not be applied directly to the glass. The use of alcohol or acetone on mounted optics is not recommended. Cleaned surfaces should air dry.

Testing the Instrument

Some operational parameters of binoculars and scopes can be tested by a user blessed with a modest degree of technical know-how and equipped with a minimum of test equipment. More complex testing must be performed by skilled technicians at the factory or professional repair facility. The tables below list parameters in each of these categories for binoculars and for scopes.

Binocular parameter	User test	Professional test
Resolution	Determine distance at which details of a dollar bill are resolved. Longer distances imply higher resolution	Read resolution chart of type shown on page 111
IPD scale reading	Measure w/scale	Measure w/scale
Diopter scale zero	None	Measure with collimator and dioptometer
Image rotation	Compare images	Compare images at infinity in collimator
Parallax at reticle	Detect large errors	Measure smaller errors
Collimation	Detect large errors	Measure dipvergence and divergence
Scope parameter		
Resolution	Same as binocular	Read chart at infinity
Diopter scale zero	None	Measure with collimator and dioptometer
Parallax at reticle	Detect large errors	Measure smaller errors

Brief descriptions of test setups and methods are given on the next two pages. Environmental testing by the user is not feasible and is costly for the factory to perform on a single unit basis. A unit suspected of these having problems should be replaced.

Test Setups and Methods

Testing can be done by the user and professional with the setups sketched here and by following the outlined methods.

Resolution tests: The binocular or scope is tripod mounted and focused on an enlarged print of a resolution target, such as the **USAF 1951**

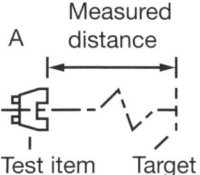

Measured
A   distance   B   Collimator

Test item   Target   Test item   Target

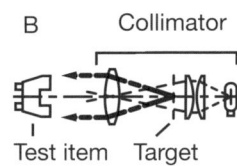

standard pattern (below left) that is located (A) by the user on a tripod at a distance typical of use of the instrument or (B) in the repair shop in the focal plane of a collimator with back-illumination. The collimator is focused for infinity by **autocollimation** from a flat mirror. The mirror is removed, and the smallest resolved target pattern is found. Resolution is calculated in lp/mm or arcmin from the target design per US MIL-STD-150A.

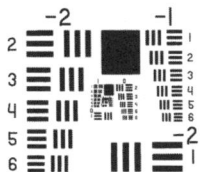

Diopter scale zero: The binocular or scope is mounted as shown in View B above and focused as sharply as possible on the center of the projected target. A **dioptometer** (see photograph right) is placed behind each eyepiece and focused on the center of the target. The defocus of the beam exiting each eyepiece is read from the dioptometer scale. It should be zero since the target is at infinity. If not, the binocular/scope diopter scale is readjusted to that reading.

Parallax at reticle: In the same setup (view A) as in the resolution test, the beam from the eyepiece containing the reticle is focused sharply on the distant target. The eye is scanned laterally across the EP. The objective or reticle (whichever is moveable) is refocused to minimize apparent motion of the reticle image relative to the scene image.

Test Setups and Methods (cont.)

In the professional test, the collimated image is observed as just described, but with a low-power microscope added behind the exit pupil to magnify the reticle pattern and the image projected upon it by the test item. This added magnification allows smaller errors to be observed in order to reduce parallax beyond that achievable by the unaided eye.

Binocular collimation, user test: A target at a distance of at least 100 m is observed sequentially through both scopes of the tripod-mounted binocular. A preferred target is a sign with vertical and horizontal crossing lines or a building window with dividing mullions in the form of a cross. The instrument is pointed so that the cross of one image is centered approximately in the field of one scope. If the cross image in the other scope appears higher or lower than the first image, dipvergence is present. If the image in that same scope appears misaligned horizontally with the first image, divergence or convergence is present. Error types can be determined by consulting the multiple-view figure on page 97. Correction of these errors requires partial disassembly of the binocular and movement of one or both eccentrically mounted objectives (see lower figure on page 97) or realignment of the erecting prisms. The latter is an extremely difficult adjustment and is best left to experts using the professional test described below.

Binocular collimation, professional test: A double test telescope (see following figure) is used to view the images seen through the binocular eyepieces in setup B on page 111. One telescope has a crosshair that is aligned with a selected target point as seen through the corresponding binocular scope. The other test telescope has orthogonal measuring scales or, preferably, a pattern of concentric circles representing different amounts of image displacement in arcminutes.

The decentration of the target image seen in this second telescope is read from the angular error markings. Both vertical error (dipvergence) and horizontal error (divergence) can be measured.

Test Setups and Methods (cont.)

Correction of the observed collimation errors can sometimes be corrected by partial disassembly and realignment of the binocular objective(s) or realignment of the erecting prisms—if that is possible in the test item.

Note that cemented prisms cannot be separated, realigned, and reassembled cheaply. In this case, it would be more economical to replace the instrument.

Binocular image rotation: If during any of the above-described collimation tests, one image appears rotated with respect to the other, the erecting prisms must be misaligned. Correction of this type of error should be attempted only by an expert technician, or preferably, the instrument should be replaced.

Professional tests of scopes: Resolution, diopter scale zero calibration, and parallax at the reticle of spotting scopes can be conducted in essentially the same ways as just described. Astronomical binoculars and scopes present some different problems because of their increased apertures and sheer bulk. For example, central obscurations require insertion of test beams within the annular scope aperture. Collimators for testing these large-aperture systems may in some cases be created from commercial astronomical telescopes.

A bit of ingenuity on the part of the test engineer can usually resolve any of these encountered problems, either at the user level or in professional test facilities.

Modular Construction

The US military night glass used during WWII and the Korean War was the 7 × 50 M17 shown on page 52. It was adapted from a Bausch & Lomb nonmilitary design. Many thousands were produced by that company and by a variety of other companies with no prior experience building optical instruments. Slightly different models, such as the Mark 28, 32, 39, and 45, were produced for a variety of US Navy applications.

The M17 Binocular featured cemented doublet objectives mounted in double eccentric rings, cast aluminum housings with leather coverings, Porro prisms attached to shelves, and IF Kellner eyepieces. Its RFOV was 7.3°, while its dimensions were ~18.3 cm long, ~21.1 cm wide, and ~7.6 cm deep. Its weight was ~1.5 kg. Each unit had >250 parts and needed ~125 special tools plus many hours of specially trained labor for depot maintenance. A failure effectively removed the unit from use for up to six months.

In 1956, a new **modular** concept was developed by the US Army in which a binocular would be developed to be optically equivalent to the M17, but with reduced weight and bulk, and capable of maintenance in the field using the absolute minimum number of components and without any special tools or skills. Prototypes were designed, built, optically and environmentally tested, redesigned, rebuilt, successfully retested, and standardized as the 7 × 50 Binocular M19 for military use.

As shown in the exploded view on the next page, the M19 comprised two identical telephoto-type objective modules of optical design similar to that shown on page 35, two identical high-performance eyepiece modules, left and right housing modules (identical aluminum castings machined as left and right modules and protected with vinyl coverings), and a hinge module that held the housings together.

With the exception of the hinge that could be disassembled to replace the eleven O-rings that provided friction in the IPD mechanism, the modules were prealigned and not maintainable.

Modular Construction (cont.)

A defective module could be unscrewed and replaced in <1 h at any reasonably clean place by any individual with minimal training. No adjustments were needed to restore performance. If a tank of dry nitrogen were available, the instrument could be **purged** of moisture after removing the two seal screws and resealed by replacing those screws.

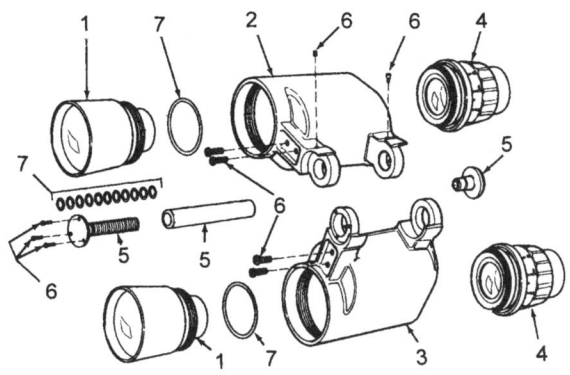

1. Objective Module, 2. Right Housing Module, 3. Left Housing Module, 4. Eyepiece Module, 5. Hinge Module (disassembled), 6. Standard Hardware, 7. Standard O-rings

Dimensions of this modular binocular were ~15.2 cm long, ~19.0 cm wide, and ~6.4 cm deep. Its weight was 0.97 kg. All of these dimensions were significantly smaller than those of the M17 Binocular. Its RFOV was the same as the M17 at 7.3°. Further, the M19's optical performance exceeded that of the replaced binocular.

Many thousands of M19s were produced, used, and field-maintained for several years. It has since been replaced by militarized versions of nonmilitary binoculars, such as the 7×50 Binocular M22 (see page 26 for the Steiner version) and the 7 × 28 Binocular M24.

General Considerations

This section compiles into tabular form much of what has been discussed earlier in this *Field Guide* regarding preferred dimensions, desirable optical parameters, environmental resistance, and operational characteristics of scopes and binoculars intended to be used in specific activities. The latter include bird watching and other nature study, hunting, military surveillance, and amateur astronomy. This information may prove helpful to those readers considering the acquisition of a new instrument for one or more of these purposes.

A brief explanation of the logic behind choice of values for key attributes is in order here. Preferences for dimensions are based on minimizing bulk and weight without unduly sacrificing optical performance [magnification, FOV, resolution, efficiency, clear eye distance (CED), etc.], environmental considerations including the ability to withstand abuse (shock, moisture penetration, abrasion resistance, etc.), and convenience of use.

Design of instruments specifically for military use in accordance with US MIL-STD-810 seems to be a lost art. Most such instruments now are adaptations of civilian designs. Moisture protection requirements more nearly follow International Ingress Protection codes IP63 or IP68, which are international standards for electrical enclosures and significantly relaxed shock requirements from the above MIL-STD. Here, drop testing four to six times from 1.2-m height is considered realistic in terms of service use. Dust and water penetration means exposure to spraying water for 5 min at 60° off vertical with volume 0.7 l/min at a pressure of 80–100 kPa.

For terrestrial applications, spotting scope and binocular magnifications as high as $60\times$ are available but rarely needed. For most applications, magnifications of $25\times$ to $40\times$ are more appropriate. Fields of view for higher magnification units are too limited for target acquisition.

Attributes for Bird-Watching Binoculars

Attributes	Types		
	Mid-size	Full-size	High power
D_{EP} (mm)	30–32	35–42	40–50
M	7–8	7–8.5	10
Min. XP (mm)	4	5	4
Min. CED (mm)	15	15	15
Min. AFOV	55°	55°	55°
Max. weight (g)	800	900	900
Environmental (dust/water)	IP63	IP63	IP63
Shock (drop height in meters/number of tests)	1.2/6	1.2/6	1.2/6
Rubber armor	Optional	Optional	Optional
Focus mechanism	CF	CF	CF
Min. focus distance (m)	3	3	4
Erecting system	Roof or Porro	Roof or Porro	Roof
Tripod mounted	No	No	Optional

Attributes for Hunting Binoculars

Attributes	Types		
	Mid-size	Full-size	Long range
D_{EP} (mm)	24–32	40–50	50–60
M	6–8	7–8.5	12–16
Min. XP (mm)	4	5	3.5
Min. CED (mm)	15	15	15
Min. AFOV	50°	50°	50°
Max. weight (g)	700	900	1200
Environmental (dust/water)	IP68	IP68	IP63
Shock (drop height in meters/number of tests)	1.2/26	1.2/26	1.2/26
Rubber armor	Yes	Yes	Optional
Focus mechanism	CF	CF	IF or CF
Min. focus distance (m)	10	10	15
Erecting system	Roof or Porro	Roof	Roof or Porro
Tripod mounted	No	Optional	Yes

Note: Two specialized types are the 8 × 56, which is used for hunting from a stand in low light, and the 10× models, which are usable when handheld for short periods, but require bracing for longer periods.

Attributes for Military Binoculars

Attributes	Types		
	Mid-size	Night glass	High power
D_{EP} (mm)	24–30	50	50
M	6–8	7	10
Min. XP (mm)	4	7	5
Min. CED (mm)	16	18	18
Min. AFOV	50°	50°	60°
Max. Weight (g)	650	1300	1500
Environmental (dust/water)	IP68	IP68	IP68
Purge ports	Yes	Yes	Yes
Shock (drop height in meters/number of tests)	1.2/26	1.2/26	1.2/26
Rubber armor	Yes	Yes	Optional
Focus mechanism	IF	IF	IF
Focus range	±4 diopters	±4 diopters	±4 diopters
Erecting system	Roof or Porro	Porro	Porro
Tripod mounted	No	No	No
Reticle	Mil. type	Mil. type	Mil. type
Laser protective filter	Yes	Yes	Yes
"Killflash" filter	Yes	Yes	Yes

Notes:

1. The reticles in military binoculars provide vertical and horizontal angular measurements in units of military mils: 1 mil = 360°/6400 or approximately 1 m at 1 km.

2. Military laser protective filters normally block the wavelengths of the ruby laser (694 nm) and YAG laser (1064 nm). The wavelength range of 800 to 850 nm is sometimes also blocked.

Attributes for Astronomical Binoculars

Attributes	Types		
	Full-size	Night glass	Mounted
D_{EP} (mm)	40–50	50–70	60–100
M	8–10	7–10	15–25
D_{XP} (mm)	5	7	4
Min. CED (mm)	15	18	15
Min. AFOV	50°	50°	50°
Max. weight (g)	1500	2100	N/A
Environmental (dust/water)	IP63	IP63	IP63
Shock (drop height in meters/number of tests)	1.2/4	1.2/4	N/A
Rubber armor	Optional	Optional	No
Focus	IF or CF	IF or CF	IF
Min. focus distance (m)	10	25	25
Erecting system	Roof or Porro	Roof or Porro	Porro
Tripod mount	Optional	Optional	Yes

Note: Weight and shock resistance are not critical for mounted binoculars.

Attributes for Spotting Scopes

Attributes	Types		
	Target shooting	Hunting	Bird watching
D_{EP} (mm)	50–80	40–80	60–80
M	20–32	15–40	15–40
Min. D_{XP} (mm)	2	2.5	2.5
Min. CED (mm)	15	18	18
Min. AFOV	50°	55°	55°
Max. weight (g)	N/A	1500	1500
Environmental (dust/water)	IP63	IP68	IP63
Shock (drop height in meters/number of tests)	N/A	1.2/26	1.2/6
Rubber armor	Optional	Yes	Optional
Min. focus distance (m)	25	10	6
45° angle eyepiece	Yes	Optional	Optional
Zoom	Optional	Yes	Yes
Digiscope capability	No	Optional	Yes

Note: Weight and shock resistance are not critical for target shooting.

Attributes for Astronomical Refractor Scopes

Attributes	Types		
	Travel/ Spotting	Visual	Photographic
D_{EP} (mm)	60–100	60–150	60–150
Min. f/number, achromat	$0.122D_{EP}$	$0.122D_{EP}$	$0.122D_{EP}$
Min. f/number, apochromat	5	6	6
Min. M	$0.14D_{EP}$	$0.14D_{EP}$	N/A
Max. M	$2D_{EP}$	$2.5D_{EP}$	N/A
Max. length (mm)	500	1500	1500
Flat field	Optional	Optional	Yes
Max. tube weight (kg)	5	20	20
Mount type	Alt/Az	Alt/Az or Equatorial	Equatorial
Tripod compatible	Yes	Optional	No

Notes:

1. Travel/spotting scopes are small and light enough to be used while traveling and on a conventional camera tripod.

2. Visual scopes are intended for visual use where a curved field can be accommodated by the user's eye.

3. Photographic scopes need large flat fields for large (up to 76.2 mm) sensors.

4. For carry-on aboard commercial aircraft, the tube length should be no more than 500 mm. The focal length may be longer if the tube is collapsible.

5. Length and weight limits also assume that a single individual is handling the tube assembly.

Attributes for Newtonian Scopes

Attributes	Types		
	Richest-Field	Visual	Photographic
D_{EP} (mm)	150–200	100–200	150–200
Min. f/number	4	6	4
Max. obscuration diameter	$0.33D_{EP}$	$0.25D_{EP}$	$0.42D_{EP}$
Min. M	$0.14D_{EP}$	$0.14D_{EP}$	N/A
Max. M	$1.5D_{EP}$	$2.5D_{EP}$	N/A
Min. unvignetted field diameter (mm)	20	12	27
Eyepiece focus tube diameter (mm)	31.75	31.75–50.8	50.8
Coma corrector	Optional	Optional	Yes
Max. tube weight (kg)	20	20	20
Max. length (mm)	1500	1500	1500
Mount type	Alt/Az or Equatorial	Alt/Az or Equatorial	Equatorial

Note: Length and weight limits assume that a single individual is handling the tube assembly.

Attributes for Catadioptric Scopes

Attributes	Types		
	Travel/ Spotting	Schmidt– Cassegrain	Maksutov– Cassegrain
D_{EP} (mm)	90–150	150–200	90–200
Min. f/number	10	10	10
Max. obscuration diameter	$0.33D_{EP}$	$0.33D_{EP}$	$0.33D_{EP}$
Min. M	$0.14D_{EP}$	$0.14D_{EP}$	N/A
Max. M	$2.5D_{EP}$	$2.5D_{EP}$	$2.5D_{EP}$
Min. unvignetted field diameter (mm)	12	25	12–25
Eyepiece focus tube diameter (mm)	31.75	31.75 or 50.8	31.75–50.8
Move primary to focus	Yes	Optional	Optional
Flat field	Optional	Optional	Optional
Max. tube weight (kg)	5	20	20
Mount type	Alt/Az or Equatorial	Alt/Az or Equatorial	Alt/Az or Equatorial

Equation Summary

Magnification (* with unity power erector):

$$M = f_{OBJ}/f_{EP}{}^*, \qquad M = \beta/\alpha, \qquad M = \tan\beta/\tan\alpha,$$
$$M = D_{EP}/D_{XP}$$

Eye pupil size:

$$\log D_{EYE} = 0.8558 - 0.0004[\log(0.3142L) + 8.1]^3$$
$$D_{EYE} = 9.08 - 0.082A + 0.00037A^3$$

Minimum object separation and maximum distance for stereo with unaided eyes:

$$\Delta L = 1000L^2\Delta\theta/B \qquad L_{MAX} = B/(1000\Delta\theta)$$

Minimum object separation and maximum distance for stereo with a binocular:

$$\Delta L = 1000L^2\Delta\theta/[(MN)(IPD)]$$
$$L_{MAX} = (N)(IPD)/[1000(\Delta\theta/M)]$$

Efficiency of a binocular or scope (in general):

$$E = R_{OPT}/R_{EYE} \qquad E = V_{OPT}/V_{EYE} \qquad E = M^{1-X}D_{EP}^X T^Z$$

Daylight efficiency of a binocular or scope (variation with scene luminance):

$$E_{Daylight} = MT^{0.25} \qquad E_{Twilight} = (MD_{EP})^{0.5}T^{0.33}$$
$$E_{Night} = (MD_{EP}/D_{Eye})^{0.5}T^{0.5}$$

Daylight efficiencies of supported and handheld binoculars:

$$E_{Handheld} = E_{Supported}/(1 + 0.05M)$$
$$E_{HandheldDaylight} = MT^{0.25}/(1 + 0.05M)$$

Equation Summary

Limiting magnitude through binocular or scope:

$$M_L = M_E + 5\log[(TD_{EP})/D_{EYE}]$$

Rayleigh resolution limit for double stars:

$$\theta_R = 138/D_{EP}$$

Dawes resolution limit for double stars:

$$\theta_D = 116/D_{EP}$$

Maximum magnification to resolve at diffraction limit without empty magnification:

$$M = 0.43 \text{ per mm of entrance pupil diameter}$$

Transmission and Strehl Ratio with central obscuration:

$$T_{OBS} = 1 - \varepsilon^2 \quad S_{OBS} = (1 - \varepsilon^2)^2$$

Minimum magnification to detect target through haze:

$$M = (L/r)(e^x) \quad x = 1.956(L - r)/R_V$$

Field stop diameter in riflescopes:

$$D_{FS} = 2(\tan\alpha)(f_{OBJ}) = 2(\tan\beta)(f_{EP})(M_E), \text{ if at } 1^{\text{st}} \text{ image}$$
$$D_{FS} = 2(\tan\alpha)(f_{OBJ})(M_E) = 2(\tan\beta)(f_{EP}), \text{ if at } 2^{\text{nd}} \text{ image}$$

Maximum real and apparent fields of scopes with interchangeable eyepieces having 31.75-mm and 50.80-mm tube diameters:

$$2\alpha_{1.25} = 1604.3/[(f_{OBJ})(M_E)] \text{ and } 2\beta_{1.25} = 1604.3/f_{EP}$$
$$2\alpha_{2.00} = 2635.6/[(f_{OBJ})(M_E)] \text{ and } 2\beta_{2.00} = 2635.6/f_{EP}$$

Equation Summary

Mass flow of water through seal:

$$W = VTR \, \frac{A \, t}{L}$$

Photography by afocal projection:

$$EFL_{AP} = M(EFL_C)$$
$$(f/number)_{AP} = EFL_{AF}/D_{EP}$$

Photography by eyepiece projection:

$$EFL_P = [EFL_{SCOPE}][S/(L-S)]$$
$$(f/number)_P = EFL_P/D_{EP}$$

Bibliography

General

Barsness, J., *Optics for the Hunter*, Safari Press Inc., Long Beach, CA (1999).

Best, B., *Binoculars and People*, Biosphere Publications, Otley, UK (2008).

Gregory, R. C., *Notes on Binoculars and Their Use*, Amwell Press, Clinton, NJ (2003).

Harrington, P. S., *Star Ware: The Amateur Astronomer's Guide to Choosing, Buying, and Using Telescopes and Accessories*, John Wiley & Sons, New York (2007).

McIntyre, T., *The Field and Stream Hunting Optics Handbook*, Lyons Press, New York (2008).

Mullaney, J., *A Buyer's and User's Guide to Astronomical Telescopes and Binoculars*, Springer, New York (2007).

Paul, H., *Binoculars and All Purpose Telescopes*, Amphoto, New York (1980).

Robinson, L. J., *Outdoor Optics*, Lyons and Burford, New York (1989).

Astronomical Use

Borgia, M., *Human Vision and the Night Sky*, Springer, New York (2006).

Crossen, C. and W. Tiron, *Binocular Astronomy*, 2nd ed., Willman-Bell, Richmond, VA (2008).

Dickinson, T., *Nightwatch*, 4th ed., Firefly Books, Buffalo, NY (2006).

Harrington, P. S., *Touring the Universe through Binoculars*, John Wiley & Sons, New York (1990).

Kambic, B., *Viewing the Constellations with Binoculars*, Springer, New York (2009).

Kitchin, C., *Telescopes and Techniques*, Springer, New York (2003).

Bibliography

Pugh, P., *The Science and Art of Using Telescopes*, Springer, New York (2009).

Mensing, S., *Star Gazing through Binoculars*, TAB Books, Inc., Blue Ridge Summit, PA (1986).

Scagell, R. and D. Frydman, *Stargazing with Binoculars*, Firefly Books, Buffalo, NY (2007).

Tonkin, S., *Binocular Astronomy*, Springer, New York (2007).

Historical

Besenmatter, W., "Recent progress in binocular design," in: *Optics and Photonics News*, Optical Society of America (Nov. 2000).

Gubas, L. J., *An Introduction to the Binoculars of Carl Zeiss Jena*, Lightning Press, Totowa, NJ (2004).

King, H. C., *The History of the Telescope*, Dover Publications, New York (1979).

Reid, W., *We're Certainly Not Afraid of Zeiss*, NMS Publishing, Edinburgh, UK (2001).

Rohan, A., *A Guide to Handheld Military Binoculars*, Optical Press, Bradbury, CA (2001).

Seeger, H., *Military Binoculars and Telescopes for Land, Air and Sea Service (Miltärische Ferngläser und Fernroher in Heer, Luftwaffe und Marine)*, Dr. Hans T. Seeger, Hamburg (1995).

Watson, F., *Stargazer: The Life and Times of the Telescope*, Da Capo Press, Cambridge, MA (2006).

Vision

Begbie, G. H., *Seeing and the Eye: An Introduction to Vision*, Anchor Books, Garden City, NY (1973).

de Groot, S. G. and J. W. Gebhard, "Pupil size as determined by adapting luminance," *J. Opt. Soc. Am.* **42(7)**, 492, (1952).

Gregory, R. L., *Eye and Brain*, McGraw-Hill, New York (1966).

Bibliography

Hecht, S., "The relation between visual acuity and illumination," *J. General Physiol.* **11**, 255, (1928).

MacRobert, A. M., "A pupil primer," *Sky & Telescope* **502**, (1992).

Overington, I., *Vision and Acquisition*, Pentech Press, London (1976).

Said, F. S. and W. S. Sawires, "Age dependence of changes in pupil diameter in the dark," *Opt. Acta* **19(5)**, 359, (1972).

Schwiegerling, J., *Field Guide to Visual and Ophthalmic Optics*, SPIE Press, Bellingham, WA (2004) [doi:10.1117/3.592975].

Southall, J. P. C., *Introduction to Physiological Optics*, Dover Publications, New York (1961).

Williamson, D., "The eye in optical systems," *Proc. SPIE* **531**, 136, (1985).

Design

Babayex, A. A. and S. A. Sukhoparov, "Design parameters for a gyroscopic stabilizer for binoculars," *Sov. J. Opt. Technol.* **39(5)**, 259, (1972).

Coleman, H. S., "Stray light in optical systems," *J. Opt. Soc. Am.* **37(6)**, 434, (1947).

Fraser, D. B., "Design of a low cost, high magnification, passively stabilized monocular, the Stedi-Eye," *Proc. SPIE* **39**, 251, (1973).

Freeman, M. H. and D. Freeman, "Innovative binocular design," *Proc. SPIE* **1780**, 711, (1993).

Giles, M. K., "Aberration tolerances for visual optical systems," *J. Opt. Soc. Am.* **67(5)**, 634, (1977).

Home, R., "Binocular summation and its implications in the collimation of binocular instruments," *Proc. SPIE* **98**, 72, (1976).

Jacobs, D. H., *Fundamentals of Optical Engineering*, McGraw-Hill, New York (1943).

König, A. and H. Köhler, *Die Fernrohre and Entfernungmesser*, Springer-Verlag, Berlin (1959).

Bibliography

Levi, L. and A. Reichert, "Roof angle error on modulation transfer function and spread function," *Appl. Opt.* **27(5)**, 915, (1988).

Mahan, A. I. and E. E. Price, "Diffraction pattern deterioration by roof prisms," *J. Opt. Soc. Am.* **40(10)**, 664, (1950).

Mahan, A. I., "Focal plane anomalies in roof prisms," *J. Opt. Soc. Am.* **35(10)**, 623, (1945).

Martin, S., "Survey of glare measurements in optical instruments," *Proc. SPIE* **274**, 288, (1981).

Mouroulis, P., *Visual Instrumentation*, McGraw-Hill, New York (1999).

Osipova, L. P., "Stabilization of handheld optical instruments," *Sov. J. Opt. Technol.* **49(2)**, 118, (1982).

Ostrovskaya, M. A., "Allowable deviations from parallelism for the optical axes of binoculars," *Sov. J. Opt. Technol.* **45**, 613, (1978).

Patrick, F. B., "Military optical instruments," in: R.Kingslake (Ed.), *Applied Optics and Optical Engineering, Vol. 5*, Academic Press, New York (1969).

Quammen et al., "Telescope Eyepiece Assembly with Static and Dynamic Bellows-Type Seal," U.S. Patent 3,246,563 (1966).

Rosendahl, G. R., "Tolerances for roof prisms," *J. Opt. Soc. Am.* **49**, 830, (1959).

Rutten, H. G. J. and M. A. M. van Venrooij, *Telescope Optics*, Willman-Bell, Inc. (2002).

Seil, K., "Progress in binocular design," *Proc. SPIE* **1533**, 48(1991) [doi:10.1117/12.48843].

Smith, W., "Techniques for First-Order Layout," *OSA Handbook of Optics*, 2nd ed., Vol. I, Part 9, "Optical design techniques," Ch. 32, McGraw-Hill, New York (1995).

Trsar, W. J., R. J. Benjamin and J. F. Casper, "Production engineer and implementation of a modular military binocular," *Proc. SPIE* **250**, 27, (1980).

Bibliography

Walker, B. H., *Optical Design for Visual Systems*, SPIE Press, Bellingham, WA (2000) [doi:10.1117/13.391324].

Wiley, R. R., "Eliminating stray light in Cassegrain telescopes," *Sky & Telescope* **232 (April)**, (1963).

Yoder, P. R. Jr., *Opto-Mechanical Systems Design*, 3rd ed., CRC Press, New York (2005).

Adjustment and Repair

MIL-STD-150A: Photographic Lenses (cancelled 2006).

Pepin, M. B., *Care of Astronomical Telescopes and Accessories*, Springer, New York (2004).

Repairing and Adjusting Binoculars, Alii Service Notes, 1996.

Seyfried, J. W., *Choosing, Using and Repairing Binoculars*, University Optics, Ann Arbor, MI (1995).

U. S. Army, *TM 9-1240-381-24&P, Binocular M19 W/E* (1978).

Yoder, P. R. Jr., "Two new lightweight military binoculars," *J. Opt. Soc. Am.* **50(5)**, (1960).

Performance

Andrews, L. C., *Field Guide to Atmospheric Optics*, SPIE Press, Bellingham, WA (2004) [doi:10.1117/3.549260].

Burton, G. J., "Transformation of visual target acquisition data between different meteorological and optical sight parameters: a simple method," *Appl. Opt.* **22(11)**, 1679, (1983).

Clark, R. N., *Visual Astronomy of the Deep Sky*, Cambridge University Press and Sky Publishing, Cambridge, UK (1990).

Coleman, H. S. and W. S. Verplanck, "A comparison of computed and experimental detection ranges of objects viewed with telescope systems from aboard ship," *J. Opt. Soc. Am.* **38(3)**, 250, (1948).

Dubenskov, V. P., A. I. Rybkina and Yu. M. Marinchenko, "The effect of vibration on the visual resolution of a telescope," *Sov. J. Opt. Tech.* **39(9)**, 522, (1972).

Bibliography

Farmer, W. M., *The Atmospheric Filter*, Vol. II, "Effects," JCD Publishing, Winter Park, FL (2001).

Hoffman, H. E., "The visibility range when observing an aircraft with and without field-glasses," *Opt. Acta* **19(5)**, 463, (1972).

Köhler, H. and R. Leinhos, "Untersuchungen zu den gestzen des fernrohrsehens," *Opt. Acta* **4(3)**, 88, (1957).

Marriott, F. H. C., "Visual acuity using binoculars," *Opt. Acta* **19**, 385, (1972).

McDowell, M. W., "Optical properties of some 8×30 binoculars," *Optik* **58(3)**, 203, (1979).

Merlitz, H., "Distortion of binoculars revisited: does the sweet spot exist?" *J. Opt. Soc. Am. A.* **27(1)**, 50, (2010).

Osipova, L. P., "A quantitative estimate of the effect of the light scattering of a viewing instrument on landscape visibility," *Sov. J. Opt. Technol.* **42(4)**, 191, (1975).

Osipova, L. P., "Viewing efficiency in telescopes," *Sov. J. Opt. Tech.* **48(4)**, 196, (1981).

Osipova, L. P., V. V. Gaykovich and S. N. Matveyeva, "The effect of light transmission of a viewing instrument on landscape visibility," *Sov. J. Opt. Technol.* **40(12)**, 733, (1973).

Osipova, L. P., "Search for objects in a sector during telescope viewing," *Sov. J. Opt. Technol.* **44(2)**, 72, (1977).

Ostrovskaya, M., "The efficiency of visual instruments over the field of view," *Sov. J. Opt. Technol.* **40(2)**, 91, (1973).

Patrick, F. B., "The efficiency of handheld binoculars," *Optik* **33**, 494, (1971).

Vukobratovich, D., "Binocular performance and design," *Proc. SPIE* **1168**, 338, (1989).

Young, A. T., "Seeing: Its cause and cure," *Astrophysical J.* **189**, 587, (1974).

Astronomical Scope Design and Performance

Bowen, I. S., "Limiting visual magnitude," *PASP* **59**, 253, (1947).

Bibliography

Garstang, R. H., "New formulae for optimum magnification and telescope limiting magnitude," *J. Royal Astron. Soc. Canada* **93**, 80, (1999).

Schaefer, B. E., "Telescopic limiting magnitudes," *PASP* **102**, 212, (1990).

Shaw, G. E., "What is a richest-field telescope," *Sky & Telescope* **192**, (1980).

Sidgwick, J. B., *Amateur Astronomers Handbook*, Faber and Faber, London (1958).

Texereau, J., *How to Make a Telescope*, 2nd ed., Willmann-Bell, Inc., Richmond, VA (1984).

Walkden, S. I., "The richest-field telescope," *Popular Astron.* **44**, 146, (1936).

Index

Index

Index

 **Paul Yoder** (BS Physics, Juniata College, 1947; MS Physics, Pennsylvania State University, 1950) began his career in optical engineering in 1948 under the guidance of Prof. David Rank in the Spectroscopy Laboratory at Penn State. He was employed for 10 years in the Optical Design Department of the U.S. Army's Frankford Arsenal and then worked for 25 years on a variety of aerospace optical systems programs at Perkin-Elmer Corp. Following retirement in 1986, he was an optical engineering consultant, largely in the development of excimer laser recontouring of the human cornea for vision correction. He is a Fellow of SPIE and the OSA, has taught many short courses in optomechanical engineering for SPIE, and authored several books on optomechanics.

 **Daniel Vukobratovich** is currently a Senior Principal Multi-Disciplinary Engineer at Raytheon Systems in Tucson, Arizona. Prior to Raytheon, he worked for fifteen years at the Optical Sciences Center, University of Arizona, where he still holds an adjunct faculty position. His primary field of interest is optomechanical design. He has authored over 50 papers, including chapters on optomechanics in standard reference works such as the *IR Handbook* and *CRC Handbook of Optomechanics*. He has taught optomechanics in 12 different countries, consulted for over 40 different companies, and holds several patents. He is a member and Fellow of SPIE, and he has received an IR-100 for work on metal matrix composite optical materials.